Books are

Aspects of
Human Biology

Theory Relevant to
Medical Laboratory Sciences

Aspects of Human Biology

Theory Relevant to Medical Laboratory Sciences

Frank Spencer F.I.M.L.T., A.I.S.T., A.R.T.

Chief Medical Technologist and Instructor, Department of Pathology, Hotel-Dieu Hospital, Windsor, Ontario. Formerly Chief Technologist, Department of Pathology, St. Bartholomew's Hospital, Rochester, Kent

London; Butterworths

ENGLAND: BUTTERWORTH & CO. (PUBLISHERS) LTD.
                LONDON: 88 Kingsway, WC2B 6AB

AUSTRALIA: BUTTERWORTH & CO. (AUSTRALIA) LTD.
                SYDNEY: 586 Pacific Highway, 2067
                MELBOURNE: 343 Little Collins Street, 3000
                BRISBANE: 240 Queen Street, 4000

CANADA: BUTTERWORTH & CO. (CANADA) LTD.
                TORONTO: 14 Curity Avenue, 374

NEW ZEALAND: BUTTERWORTH & CO. (NEW ZEALAND) LTD.
                WELLINGTON: 26–28 Waring Taylor Street, 1

SOUTH AFRICA: BUTTERWORTH & CO. (SOUTH AFRICA) (PTY) LTD.
                DURBAN: 152–154 Gale Street

Suggested U.D.C. No. 575/577
Suggested Additional Nos. 576.1/.3 and 612.01

I.S.B.N.   0 407 70400 0

D
574. 0246'1
SPE

Printed in Great Britain by
Redwood Press Limited
Trowbridge, Wiltshire

To
My Wife

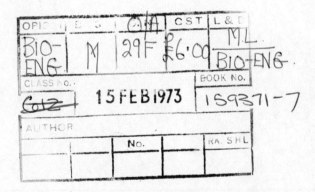

Contents

Preface

Primarily this work was conceived as a logical extension of the text of *Introduction to Human and Molecular Biology*, published by Butterworths in 1970. The thesis of this introductory work was the general concept of Man as a multicellular organism and, as such, the major portion of the book was devoted to the physiology of those systems constituting the total organism. However with this text the approach has been narrowed to a 'particulate' consideration of cell physiology, and those topics related to the maintenance of the *status quo* of the cell with the environment.

The book commences with two chapters which are concerned with the origins and development of terrestrial life and with an appreciation of the basic unit of life — considered against a background of the historical emergence of the cell theory — the aim here being to enable the reader to view the whole in perspective. In subsequent chapters consideration is given to macromolecular development, from those basic raw materials present in the biosphere which are required to sustain normal heterotrophic cellular activity. The final chapters deal with the principal biomechanisms involved in the maintenance of somatic integrity.

The contents should be of interest to those students reading HNC subject courses in Medical Laboratory Sciences, and for Special Fellowship examinations of the Institute of Medical Laboratory Technology — though it is considered the book should prove useful also to undergraduates taking parabiological courses.

I would like to express my sincere thanks to Dr. Trevena Hyde — Pathologist at Hotel-Dieu Hospital, Windsor, Ontario, and Adjunct Professor of Biochemistry at the University of Windsor — who is author of the final section on Homoeostatic Function. In addition to thanking Dr. Hyde for his invaluable contribution I would like to express my appreciation to the Staff of Butterworths Medical Division who have at all times been most helpful and understanding during the difficult and sometimes painful process of editing — a task which was somewhat frustrated by my sudden move to Canada.

Windsor, Ontario   FRANK SPENCER

ix

Part I
Molecular and Cellular Organization in the Biosphere

1— A General Survey of the Origins and Development of Terrestrial Life

(The origins of life – The chronology of life – Hominid evolution)

THE ORIGINS OF LIFE

In the nineteenth century the Aristotlean concept of spontaneous generation was exposed by the work of Louis Pasteur as being inconsistent with the 'Age of Scientific Materialism'. In historical context the fusion of the ideas of Pasteur and Darwin appeared to finalize the prospect of ever solving the problem of biogenesis on Earth. But though Pasteur's brilliant experimentation showed quite categorically that life did not arise from non-living material – at least, not under the conditions which he had employed – we are nevertheless required by logic, if we are to accept the precepts of Darwinian evolution, to assume that such an initial genesis did indeed precede the development of the contemporary living world.

Hence, by the end of the first quarter of the twentieth century there had been a revival – albeit a disguised one – of the classical concept of generation. Both Oparin (1924) and Haldane (1929) published papers independently postulating that the development of the living organism was preceded by a period of chemical evolution. Initially both papers were regarded as wildly speculative, but subsequent biochemical research has shown there to be a prima facie case for supposing that life can arise from indigenous inorganic material. In fact the problem is no longer one of whether it can happen – but of exactly how it did happen.

In company with other workers Miller (1953), and later Oro and Kimball (1961 and 1962), have sought to offer explanations of conjectured chemical pathways leading to the origins of pre-biotic systems, giving rise subsequently

3

to life on earth. The net result of such work viewed against the backcloth of existing knowledge in the parabiological sciences indicates that life must have originated as the direct result of definitive biochemical evolution.

On the basis of geological evidence it is assumed that biogenesis occurred in the Pre-Cambrian era some 4×10^9 years ago (Nagy and Nagy, 1969), and that the continuing biopoiesis of the biota has depended largely on several important and seemingly accidental factors. Thus, the earth possessed a tilt in its rotational axis away from normal to the plane of revolution about the sun, giving rise to the seasonal changes, and the important ecological variations encountered in latitude environment. Coupled with this was the fact that the planet had originated and positioned itself in an elliptical and bio-optimal solar orbit, which permitted the evolution of a definitive atmosphere and temperature, this resulting in the presence of water in the liquid state extending over the majority of the surface of the earth.

Initially the earth's primitive atmosphere contained mixtures of hydrogen, methane and ammonia which were acted upon by electrical discharges in the presence of water vapour, ultra-violet light and heat, this resulting in the synthesis of complex organic molecules. The determining environmental condition during this crucial period was an anaerobic and strongly reducing atmosphere. Free oxygen appeared in the earth's atmosphere at a much later period — arising, in fact, as a consequence of the biological evolution of photosynthesis. With some pedantic reservations authorities generally accept that, at this stage within the primitive oceans, chemical pathways were established resulting in an accumulation of organic compounds, often referred to as the primitive soup (Oparin, 1957).

The work of Miller and Urey (1959), and of others has shown that on the occurrence of stable amino-acid compounds in this primitive soup, under reduced conditions, they will eventually become reactants — leading to the formation of nucleotides, such as adenine (Oro and Kimball, 1961) or uracil (Fox and Harada, 1960). Matthews and Moser (1966) have shown that when the condensation products of a mixture of hydrogen cyanide and ammonia are hydrolysed no fewer than fourteen amino acids are produced — the initial products of this hydrolysation being polypeptides! Thus, with the selection of improved catalysts occurring within the primeval soup, polymers and the simple analogues of the proteins and nucleic acids evolved, producing molecular aggregations with characteristics closely resembling and associated with proto-living processes.

In the coacervate hypothesis Oparin (1938) attempts to explain the transition mechanism of molecular structures required to achieve threshold cellular organization — though workers such as Bernal (1967) argue that coacervation occurred much later in the chemical evolutionary scale than Oparin proposed.

Coacervates are proteins or protein-like structures held together by minute droplets of water commandeered from the surrounding environment. This is achieved by virtue of the fact that when proteins are dissolved in water, a portion of the molecule becomes ionized and so attracts molecules of water, this resulting in the protein molecule becoming surrounded by an organized layer of water. Hence, protein complexes will become concentrated from aqueous solution into localized areas of higher molecular concentration.

Bernal (1961) conjectures that coacervate structures were preceded by the polymerization of macromolecules such as those of the proteins and nucleic acids, occurring as the direct result of the accumulation of organic flotsam on beach mud and estuarine clays. Recently it has been suggested that the pools surrounding the alkaline springs which were commonplace on primeval earth may have been one of the sites where life arose.

With the establishment of biochemical mechanisms, whereby molecular replication was achieved, escalation in biomolecular organization was inevitable. Thus, there occurred a subtle metamorphosis, following a sequential scheme of natural selection, from the inanimate molecular state to a structural unit with the potential of life: a unit, that is, possessing an inherent ability to manufacture a self-replicating polymer enclosed within a semipermeable membrane, and capable of adjusting, spontaneously, its own internal processes to external conditions, thereby maintaining systemic coherence.

The interesting feature of this modern concept of the spontaneous generation hypothesis, is that at first sight life appears to be a phenomenon isolated in space and time − with the demarcations between the living and the non-living blurred, making it extremely difficult to arrive at an exact definition of the boundary between the two states. Objectively, life appears to be a property possessed by a structure of particular molecular architecture − that is, an organization of numerous different substances having specific inter-spatial relationships which collectively possess an intrinsic principle of purposeful integration and ability to work upon 'self'; and which moreover, for a certain period of time, is able to manifest this tangible property in its utilization of matter. Non-living chemical processes occur with an available supply of free energy that is purely fortuitous, and independent of the pathways it subsequently activates − whereas a living system actively perpetuates energy supplies which are required for the reproduction and maintenance of the total integrity of the system. Oparin (1957), has this to say on the distinction between living and non-living systems: '. . . . living things differ fundamentally from all such (physico-chemical) open systems in the orderly regulation of their metabolism and the purposefulness of their internal structure. Not only are the many tens and hundreds of thousands of chemical reactions which occur in protoplasm, and which together constitute metabolism, strictly co-ordinated with one another in time, harmoniously composed into a single series of processes which constantly repeat themselves, but the whole series is directed towards a single goal, towards the uninterrupted self-preservation and self-reproduction of the living system as a whole in accordance with the conditions of the surrounding medium'. Consequently any system falling short of the latter definition fails in his view to qualify as an example of life.

TABLE 1.1

A Summary of Precellular and Cellular Evolution

The determining environment and location of life, or prebiotic forms	Energy source	Evolutionary product
Anaerobic. Strongly reducing	Ultra-violet light	Simple compounds
Atmosphere: containing simple molecules of hydrogen, nitrogen, carbon and oxygen	Heat and electrical discharge	High-energy phosphates – proto-enzymes
Anaerobic	Heat and electrical discharge	Accumulation of organic molecules in primeval oceans – of amino acids, acetates, uracil and adenine, for example
Estuarine muds and proto-soils	Local heat – ? Radioactivity (ultra-violet photo-association)	
Ozone layer near planet surface	:	Polymerization of macromolecules With natural selection: complex organic substances such as carotenes, proteins and nucleotides
Mainly anaerobic, with traces of oxygen under surfaces of pools		
Mainly anaerobic, with traces of oxygen under surfaces of pools	:	Coacervates and formation of prototype membranes
Oxygen concentrations increasing	Ultra-violet light via photo-reduction. Visible light	First appearance of procaryotic cells. i.e. the bacteria. Photophosphorylation. Metabolic units replicating at the expense of accumulated organic food in previous stages of biochemical evolution. Hence rapidly diminishing supply of these original food stores
Mainly aerobic with anaerobic pockets Formation of oceans – commencement of the Continental drift	Fermentation	
More carbon dioxide	Visible light via photosynthesis	Photo-reduction of CO_2 at expense of organic and inorganic hydrogen donors. Autotrophic plants. New food supply. Evolution of respiring organisms
Less carbon dioxide	Respiration	Eucaryotic cells – single and multicellular. Evolution of plants and animal phyla. Plants and animals now completely dependent on photosynthesis with free carbon dioxide
Atmosphere comparable to contemporary in oxygen content		Equilibration to present conditions. Continuous turnover of a nearly constant volume of organic matter

Life appears to be a distinctive and paradoxical manifestation of matter, simply because of the difficulty encountered in visualizing the complex dynamics of the evolutionary process in their immensity. Each major step in the evolutionary process involved an increase in complexity of organization, which has ultimately led to the dominance of a new structural design and, reciprocally, the elimination of the less adapted forms from the changing environment. Thus, with the development of the first successful cell system there followed a natural displacement and cannibalization of those forms embodying a primitive type of organization (pre-biotic systems). This probably explains why pre-biotic systems have not been demonstrated, and also accounts for the apparent monophyletic origin of the contemporary living world. During the inanimate period of chemical evolution the question of chirality — that is, rotation of a plane of light to right and left — did not arise. In nature amino acids can exist in two mirror image forms, but without exception all contemporary molecular configurations are constructed from L-amino acids, and not from the equally likely D-amino acids. Hence with the advent of living systems a pattern was assumed — either by chance or by predetermined electron states as yet not understood — whereby the chirality of subsequent molecular configurations ultimately to be incorporated into the structure of the living system was compatible (Kvenvolden, Peterson and Pollock, 1969).

The subsequent dissemination of cellular organisms on the planet's surface resulted in the gradual depletion of those organic compounds accumulated during the preceding periods of biochemical evolution — and also in concomitant changes in the earth's atmosphere which eventually prevented the synthesis of organic compounds by photochemical means. Thus a photosynthetic process was established, which is still the ultimate energy source in the contemporary biosphere. At this point life acquired the potential for a balanced non-exploitive economy, and thus those conditions at present prevailing on earth were slowly established. An endeavour is made in Table 1.1 to summarize the current view of the succession of pre-cellular and cellular evolution.

THE CHRONOLOGY OF LIFE

Structurally the earth consists of a number of concentric shells of different materials arranged in order of increasing density. The outermost shell is a thick rigid crust composed of solidified granite and basalts resting on a more plastic mantle composed of basalt with heavier rocks beneath. The mantle surrounds a soft core of pure iron.

Four-fifths of the earth's crust lies submerged beneath water, which has filled a deep system of rifts extending down the centre of all the ocean beds and circumventing the continent of Antarctica. Recent geophysical studies of the earth's crust have revealed that these rifts in the ocean beds are slowly drifting apart, resulting in the gradual upward displacement of the

Himalayan, Andean and Alpine mountain systems of the continental land masses. Furthermore, this slowly emerging picture of the continental drift, supports the old Wegener hypothesis expounded in 1912 (*see* Du Toit, 1937), which supposed that the now separated land masses of the southern continents originally constituted a unified whole known as Gondwanaland. A reconstruction of Gondwanaland – using a computer program to match coastal contours, combined with geological and geophysical data – was recently published (Smith and Hallam, 1970); in this a part of Antarctica was matched with a complementary morphology on the East African coastline, India and Australia. Similarly, a geometric 'fit' between West Africa and South America was made, thus completing the picture of a continuous continental area. In their discussion of the disintegration and movement of the continents towards their present configuration, Smith and Hallam are of the opinion that the initial rifting occurred in the Jurassic period, with much of the dispersal taking place in Upper Cretaceous and Tertiary times.

Though an accurate time-table for the origin of life on earth has yet to be drawn up, it is known with reasonable accuracy that the earth was formed in consequence of some unknown cataclysmic event occurring in the cosmos some 5×10^9 years ago. This date has been determined by studying the relative quantities of disintegration products of radioactive elements trapped within the sedimentary rock layers of the lithosphere, sampled at varying depths. Such elements as uranium and thorium possess this extraordinary property of atomic instability, in which the gradual disintegration is characterized by the intermittent emission of high-velocity α particles. Hence by the loss of their atomic constituents, uranium and thorium, for example, are gradually converted by their own radioactivity into helium and lead. The rate at which this conversion occurs is low, but can be measured by counting the number of α particles emitted from a given rock mass over a measured period of time. With this data, and by assaying the amount of lead or helium produced by radioactive decay, it is possible to date those rocks possessing such elements. Because the rate of radioactive decay is independent of temperature and pressure, or any other physical or chemical variable, the use of such a method of measurement is valid, since it satisfies all the necessary requirements of accuracy. Accordingly, by dating the oldest rocks known on earth – those which are to be found in the Canadian, Scandinavian and South African Shields – it is known that the formation of a stable crust took place some 3×10^9 years ago (Stubbs, 1965), which is regarded as the point of departure for biochemical evolution. This was followed by an estimated five hundred million years before the first cell system appeared.

The earth's crust consists of disunited fragments of overlapping sedimentary rock layers forming what is known as the geological column. Rocks can be broadly divided into three major classes: igneous rocks, which have solidified from a molten state; sedimentary rocks which were deposited on the crust's surface, not originally in a molten state; and metamorphic rocks which may have been either igneous or sedimentary, but have since lost their original characteristic features. It is in the sedimentary rock layers that fossils are found which, apart from indicating the route of historical geology, also represent a direct palaeontological record of the course taken by biological evolution. There are numerous kinds of sedimentary rocks, of which the commonest by far are the shales. The shales are hardened clays; other such rocks are sandstones and limestones, all of which intergrade. It is because of this property of intergradation that sedimentary rocks present a classification problem geologically.

Generally speaking the term fossil refers to the preserved remains of living things — these providing direct evidence of their former existence, their forms immobilized in time and space as were the human larval images at Herculaneum. Such evidence often comes in the form of an impression, or print, of an animal or plant structure found on a rock surface. Similarly the negative image of a structure may be preserved in the form of a mould — or more directly, the morphology of the living structure may be preserved simply in the petrified state.

The Pre-Cambrian Era

The whole of geological time can be divided into five eras known as the Archaeozoic, Proterozoic, Palaeozoic, Mesozoic and Cainozoic. The Archaeozoic and Proterozoic are collectively termed the Pre-Cambrian era. As implied earlier in the text there is no direct evidence of the origins of life — our understanding of the Beginning is simply inferential and based on experimental evidence. A clear and continuous fossil record begins in the Upper Proterozoic era. Recent finds in the Ediacara Hills of South Australia have yielded a wide variety of Pre-Cambrian fossils, including several types of jelly-fish, corals and two other forms that resemble no other known form of organism (Glaessner, 1961). The Pre-Cambrian era is thus characterized by sparse fossil evidence of the earliest forms known to exist — that is, plants related to the Thallophyta; animals such as the Coelenterata and the Trilobita. The trilobites are a group of invertebrates belonging to the Class Arthropoda, and superficially resembling the woodlice. They evolved rapidly during the Cambrian and eventually became extinct by the close of the Palaeozoic era. All of which clearly indicates a wide variety of biota that had already differentiated into plants and animals — implying a long and, as yet still unknown, evolutionary past.

Very little is known about terrestrial conditions during the Upper Proterozoic and early Cambrian periods, but it is generally agreed that no significant invasion of the land by living organisms occurred during this time, and that it was probably a phase when life existed exclusively in aquatic environments.

The Palaeozoic Era

The Proterozoic era was followed by the Palaeozoic which spanned a major period of time – more than three hundred million years. The Palaeozoic era is subdivided into six major periods, the early periods being termed the Cambrian, Ordovician and Silurian, and the later periods the Devonian, Carboniferous and Permian. North American palaeontologists divide the Carboniferous into the Mississipian and Pennsylvanian periods respectively. The early Palaeozoic era is characterized by an abundance of fossil evidence indicating a rapid proliferation in the organizational diversity of the flora and fauna of the biosphere. It is from this era that our principal evidence regarding the subsequent emergence of major biological groups appears. By simple extrapolation, it is possible with evidence of this and of later periods to formulate a working hypothesis regarding the origins and development of such biological groups.

By the end of the Cambrian period all the major groups of the invertebrates were in existence. The earliest signs of vertebrate development are to be found in the rocks of the Ordovician period. Fossil evidence indicates that by Silurian times plants in ever-increasing numbers, were inhabiting the land, where they rapidly evolved. The Devonian period provides the first evidence of vertebrates emerging from their aquatic environments and living out part of their life cycle on land as amphibians. Also during the Devonian and Carboniferous periods development of the terrestrial flora was represented by the huge and dense tropical forests covering large areas of the land mass. The net result of this was the eventual deposition, over vast areas, of fossilized biological fuels which were utilized some 250 million years later as a source of energy to power a developing industrial revolution occurring in Human society. It has been suggested that these fuel deposits were the product of an incomplete evolution of the terrestrial biota – that is to say, that animals and other heterotrophs had not successfully perfected the transition from an aquatic to a land existence. The Palaeozoic era ended with the development of the reptiles, and another geological revolution resulting in the wholesale extermination of many animal and plant forms under the extreme pressure of competition and forces of natural selection – such that for every plant species that had become extinct as a result of the geological change a whole spectrum of such specialized dependents as herbivores, predators and parasites became extinct in turn.

The Palaeozoic era was an important one for a variety of reasons, but probably the most significant in relation to the evolution of the vertebrates was the development of the amniote egg – a necessary antecedent to a successful and permanent terrestrial conquest. Fundamentally the major limitation of the amphibian design was that for their reproduction (involving external fertilization of the female eggs) and the early development of the embryo an aquatic environment was essential. As a result, the amphibians were faced with several problems related to their biological design, namely: (1) the hazard of dehydration, if exposed to the air and, (2) the effect of gravity — which is experienced with greater force on land than in the water. To overcome problem (1) the amphibians would require radical changes to the design of their eggs and mode of reproduction; whereas problem (2) would involve structural modifications to the animal's supporting skeleton. In overcoming the initial problem the reptilian design utilized the amniote egg (involving internal fertilization), which is protected by a hard shell, thus permitting it to be laid on land. Furthermore the amniote egg is more richly endowed, than its amphibian counterpart, with a nutritious yolk — enabling the embryo to attain a considerable degree of development prior to birth. Thus the reptilian embryo develops within the protection of a shell, chorion and amnion. The amniotic sac, which contains the embryo, is in fact a miniature replica of its ancestral pond. By contrast the amphibian has to seek its food as an essentially fish-like larva in a hostile external environment, and only after considerable time is it released from the water, by a graded metamorphosis involving atrophy of the gills and expansion of the lungs followed by reduction of fish traits and development of limbs suited to subsequent terrestrial locomotion. Thus, as a result of subtle adaptation, the reptiles were released from an aquatic life and the adult was no longer compelled to return to its natal element for reproductive purposes.

The Mesozoic Era

During the next 120 million years known as the Mesozoic era, the reptiles reached the zenith of their development, this being characterized by the appearance of such spectacular creatures as the dinosaurs (for example, the brontosaurus and tyrannosaurus) and the pterosaurs, the flying reptiles — precursors of the avian species. A recent reconsideration of brontosaur remains suggests that the large dinosaurs were in fact fully terrestrial and not semi-aquatic as generally suggested in museums and in published works (Bakker, 1971). It is also considered that the brontosaurs had an enormous impact on the terrestrial ecology, in as much as they must have opened up thick forest, and kept the undergrowth from becoming dense, in much the same way as elephants do in Africa today.

The Mesozoic era is divided into three distinct periods: Triassic, Jurassic and Cretaceous. The middle Mesozoic period gave rise to an important development in the community of the terrestrial flora — the appearance of Angiospermae, or flowering plants. This was a key event in the evolution of the earth's vegetation, leading to the establishment of the modern characteristics of the biosphere. Loren Eiseley (1957), in his book *The Immense Journey* describes beautifully the impact which the flowering plants had on the environment: '. . . . the true flowering plants (angiosperm itself means 'encased seed') grew a seed in the heart of a flower, a seed whose development was initiated by a fertilizing pollen grain independent of outside moisture. But the seed, unlike the developing spore, is already a fully equipped embryonic plant packed in a little enclosed box full of nutritious food . . . . In a movement that was almost instantaneous, geologically speaking, the angiosperms had taken over the world. Grass was beginning to cover the bare earth until, today, there are over six thousand species. All kinds of vines and bushes squirmed and writhed under new trees with flying seeds.

The explosion was having its effect on animal life also. Specialised groups of insects were arising to feed on the new sources of food and incidentally unknowingly, to pollinate the plant . . . . A new world had opened out for the warm-blooded mammals. Great herbivores, like the mammoths, horses and bisons appeared. Skulking about them had arisen savage flesh-feeding carnivores like the now extinct dire wolves and the sabre-toothed tiger. Flesh eating though these creatures were, they were being sustained on nutritious grasses one step removed . . . .'

During the Mesozoic era there is evidence to support an equally climactic event in the fauna community — namely, the emergence of the first mammals. Skeletal evidence of Mammalia during the Jurassic is not abundant, but suggests that the forms were undoubtedly mammalian but had retained a number of reptilian traits. These Jurassic reptile-like mammals, known as Pantotheria, during the Cretaceous period gave rise to the Metatheria (marsupials) and Eutheria, or placental mammals. These mammalian types will be described later in the text. It is considered that the earliest placental mammals were insectivores which appeared during the Upper Cretaceous, and evolved rapidly. Apart from giving rise to the lagomorphs (hares and rabbits) and rodents, the insectivores also provided the ancestral basis for the primates.

The Mesozoic era ended not only with the extinction of the large 'ruling' reptiles, but also with the massive extermination of many other animals and plant forms. The Age of Reptiles was followed by the Cainozoic era which spanned an estimated 70 million years, and is often referred to as the Age of Mammals. During this era, there was no radical alteration in the pattern of organizational development of the aquatic vertebrates; on land, though, dramatic and significant transformations were occurring — the mammals assuming variety and ascendency over all other classes of animals.

Throughout biological history with every major breakthrough in design there follows a reciprocal expansion of this new organizational pattern whereby the intrinsic potentialities of the new design are fully exploited. Diversification, which is the basis of speciation, is fundamentally an experiment in adaptation and specialization, whereby successful coexistence of forms occupying the same time and space is achieved by their differential utilization of the resources of the environment — thus reducing competition. However specialization can endanger or limit a particular design, particularly if it is subject to a serious change in environmental conditions and is unable to cope with the transition.

Mammalia represent the apex of a dynamic and sequential biometamorphic scheme through which preceding specie designs have evolved as a product of selective adaptation. Thus, the amphibians evolved from fish, so giving rise to the reptilian organizational designs, and from these the mammalian and avian patterns evolved in turn.

The mammals, like the reptiles and birds, are differentiated in organizational design from the amphibians and fish by the possession during the early stages of their life-cycle of an amnion and allantois, and by the absence of external branchiae or gills. On the other hand, the mammals and birds differ from the reptiles in being warm-blooded, and having a heart with four chambers and a complete double circulatory system. Further comparative differences with the reptiles and birds are that the mature circulating mammalian erythrocyte is non-nucleated; the lungs are suspended in a thoracic cavity separated from the abdomen by a muscular partition, the diaphragm, which figures principally in the mechanics of respiration; the heart is also contained in the thoracic cavity; the skin is more or less covered with hairs; the skull forms a single unit, composed of bone plates fused together − to which is articulated the lower jaw composed of two halves (rami) joined at the chin; the dentition is heterodont − that is, characterized by different kinds of teeth with different shapes and functions − as opposed to the homodont dentition of reptiles, in which all the teeth have the same general appearance and function; the skull is articulated with the vertebral column by means of two condyles which lock into the ring-like first cervical process (atlas); the second cervical vertebra (axis) possesses an odontoid process on which the head and atlas can rotate.

In all mammals reproduction is bisexual and they conceive their young within the maternal uterus and, after the offspring are born, nourish them by means of mammary glands. However, there are three basic variations in this process, in mammals. They occur respectively in:

Prototheria mammals. − These produce their young in a casing usually referred to as an egg, from which they hatch. Examples of this type are the monotremes of Australasia (the duck-bill platypus).

Metatheria mammals. − The parent in this instance bears the young in an immature condition in a pouch equipped with mammae − a characteristic of the marsupials.

Eutheria mammals. − Here the placenta enables the young to remain *in utero* for a greater period of time until an advanced state of development is reached.

Nourishment of mammalian offspring post-partum is generally achieved by a fluid secreted by glands on the ventral side of the female. However the prototheria lack such mammary glands in the general sense of the word. The platypus for example feeds its young via an epidermal secretion. Thus the possession of

13

mammary glands *per se* should not be used as a criterion for the classification of mammals.

The Cainozoic Era

As previously mentioned the Mesozoic era was succeeded some 70 million years ago by the Cainozoic. This era is divided into two periods – Tertiary and Quaternary. The Tertiary has been subdivided into the following epochs: Palaeocene, Eocene, Oligocene, Miocene and Pliocene. The Quaternary which commenced approximately two million years ago has been divided into two epochs namely, the Pleistocene and Holocene (or Recent). It is generally considered that there have been four major ice ages during the Pleistocene, designated Gunz, Mindel, Riss and Wurm glacial periods respectively.

By the end of the Palaeocene epoch all the major groups of mammals were in existence. With the mass extinction of the large reptiles, the evolutionary potential of Mammalia could be exploited; this, as already implied, resulted in the rapid proliferation and diversification of the group. A product of this evolutionary explosion was the Order of Primates to which Man belongs. For clarity the Major subdivisions of the primate order are summarized in Table 1.2. The order was defined in 1873 by Mivart, as follows: 'Unguiculate, claviculate, placental mammals, with orbits encircled by bone; three kinds of teeth, at least at one time of life; brain always with a posterior lobe and calcarine fissure; the innermost digits of at least one pair of extremities opposable; hallux with a flat nail or none; a well-developed caecum; penis pendulous; testes scrotal; always two pectoral mammae'. It will be seen from Table 1.2 that the order is subdivided into two suborders – the Prosimii and the Anthropoidea.

The Prosimii include four superfamilies: Lemuroidea, Lorisoidea, Tarsioidea and Tupaioidea. However it should be noted here that some zoological taxonomists may exclude Tupaioidea (the tree-shrews) from the suborder Prosimii: others might assign them to the Insectivore – considered by many to be a 'taxonomic-dustbin' of primitive placental mammals. The reason for this apparent controversy is simply that the group is in biological limbo, representing an apparently basal form transitional with Insectivora. From a general point of view the prosimians are quadrupeds which are well adapted to an arboreal life and with a fairly well developed visual sensory system. Although the prosimians rely more on visual than on olfactory information, their sense of vision is not as finely developed as that of the anthropoids.

Though this is not clearly documented in the fossil record, it would appear that the primates evolved from insectivore ancestors in the Upper Mesozoic era. Essentially the insectivores are terrestrial, and are largely dependent on a sense of smell to inform them about features of their environment – whereas the primates are fundamentally visually orientated. The first insectivore–primates had to adapt to a completely new world, full of new problems and how this

14

transition was successfully accomplished is of fundamental importance to an understanding of the emergence of Man. A study of contemporary prosimians provides us with an insight into the trends and changes demanded of a biological design by an arboreal adaptation.

TABLE 1.2

Major Subdivisions of Primates

Suborder	Infraorder	Superfamily	Family
Prosimii (lower primates)	Lemuriformes (lemurs) Lorisiformes (lorises) Tarsiiformes (tarsiers) Tupaiiformes (tree-shrews)		
Anthropoidea (higher primates)	Catarrhini (Old World primates)	Cercopithecoidea (Old World monkeys) Hominoidea (hominoids)*	Pongidae (apes, pongines) Hominidae (hominids)*
	Platyrrhini (New World primates)	Ceboidea (New World monkeys)	

*The terms hominid and hominoid should not be confused. Hominoid is a colloquial term for the superfamily Hominoidea which includes Human and anthropoid ape families. A hominid is a member of the zoological family Hominidae, to which all forms of man, living or extinct, are assigned.

As already implied, the modern Tupaia, the tree-shrew, represents a transitional form. As a prosimian it has retained many features characteristic of the insectivore — namely, a continuing reliance on the olfactory senses for survival, though comparison with a eu-insectivore reveals this feature to be of less importance, since in common with the primate, a premium has been placed on the development of the visual capacities. Thus adaptation to an arboreal existence has established distinctive trends of evolution, which have been followed in the development of different primate groups (see Clark, 1955, 1965). The tendencies which characterize the Order Primates involve: (1) limb structure and (2) cranial organs.

15

The Order of Primates

Limb Structure

In the primate there has been a perpetuation of the primitive mammalian organization which has involved (a) keeping well developed clavicles, to permit greater flexibility of the pectoral appendages and (b) retention of the primitive vertebrate pentadactyl appendages.

The possession of a prehensile hand or foot is of paramount importance to survival in a forest canopy. Naturally the differentiation of these primitive terrestrial vertebrate characteristics was slow — the forelimbs, and in particular the hands, gradually becoming more actively implicated in the acts of grasping, balancing and directing the general progress of the early primates as they sampled their new habitat. It is interesting to note here that all primates are born with a clinging reflex. A primate infant must have the facility to hang on to its mother as she moves through the trees; if it were unable to do so, then it would perish.

Cranial Organs

The arboreal environment is full of unpredictable changes of state. In moving from branch to branch the primate required accurate visual information — namely, the stereoscopic vision which permits depth perception. Hence the visual centres of the primate cortex expanded to deal with this increased amount of information; a reciprocating reduction followed in the development of the olfactory centres and related organs which are of less importance to an arboreal creature. Also essential to the prosimians was an increasing sophistication of those areas controlling motor dexterity — namely the neopallium.

The overall implication here is that in the insectivore–primate transition, the brain underwent substantial change as a result of adaptation to life in the tree. Table 1.3 summarizes the general characteristics of the primate.

The early prosimians were a highly successful group, and appear to have been widely distributed throughout the Old and New World. However their numbers slowly diminished, such that by the close of the Eocene epoch, they had become extinct in North America and Europe. One reason given for their decline is that the group experienced intense competition from the emerging Anthropoidea.

In the New World, the Ceboidea evolved from a prosimian base, and in the Old World some prosimian, perhaps Tarsioidea, gave rise to the Cercopithecoidea and Hominoidea. Recent immunological evidence from living primates, related to our present interpretation of the primate fossil record, suggests that divergence between the Hominoidea and Cercopithecoidea occurred during the Oligocene–Miocene epochs some 30 million years ago; and divergence between the Hominidae and Pongidae (see Table 1.2) in the Upper Pliocene epoch, approximately 5 million years ago (Sarich, 1967). Prior to this time Hominoidea had shared a common ancestry, namely the Dryopithecinae, which existed in a number of forms in Asia, Europe and Africa during mid-Tertiary times. These are represented in the fossil record by fossilized jaws and teeth. The lower molar teeth of this group possess what is called the Y-5 or Dryopithecus pattern, which is found only in the hominids. Recent finds in Africa and India, from late Miocene— Pliocene sites, have revealed a more advanced Dryopithecine known as

TABLE 1.3
A Summary of General Primate Characteristics

Modifications to basic primitive mammalian organization	Functional product
Retention of clavicle Retention of prehensile extremities	Grasping-climbing-brachiating
Nails replace claws	Development of pads well endowed with tactile sensory tissue — permitting delicate manipulations
Expansion of cerebral cortex	(1) Development of stereoscopic vision (2) Reduction of muzzle, and related olfactory dependence
Generalized dentition	Omnivorous diet
Improved foetal membranes	Increased gestation period
Large neopallium, and increased period of infant dependence	Capacity for behaviour-modification — 'protoculture'

Ramapithecus. Several palaeoanthropologists have speculated that these pre-hominid types were bipedal and tool users! However it must be remarked that statements as to whether or not Ramapithecus was a tool maker or user, or indeed any assessments of their ancestral or collateral status are highly speculative and tentative. For those interested in looking at the evidence references useful as a baseline are Leakey, 1968a and 1968b.

The Miocene—Pliocene juncture is clearly of special significance in the higher primate evolution — but as yet we know very little about the ancestral forms and the forces motivating their shift from an arboreal to a terrestrial habitat. The net result, however, was that the first savannah bipeds found themselves biologically ill-equipped to maintain the *status quo* in a new open environment — living as they did, under the constant threat of predators and adverse conditions generally. The success of these early hominids points, therefore, to the likelihood of their use of culture as a major means of adaptation. This distinctive capacity for culture in hominids is seen to be dependent fundamentally on the maximum development of those innate primate characteristics outlined earlier — rather than on the evolution of new specialized anatomical structures. Thus the hominid condition is such that it permits the transmission of acquired skills and knowledge by learning, and as a consequence is no longer able to survive without this capacity. Hence these primitive bipeds stood on the threshold of humanity, and possessed what might be termed 'protoculture'. The palaeontological evidence for the existence of culture, from the Present to the Lower Pleistocene epoch, suggests a single tradition. At present, anthropologists recognize four distinct stages in human evolution; Australopithecus, *Homo erectus* (Pithecanthropines), *Homo sapiens neanderthalensis* and *Homo sapiens sapiens.*

It became apparent during recent years that much of the controversy surrounding discussions of fossil man were founded on misconceptions of well known taxonomic and palaeontological principles. Indeed, it is probable that the one single factor, above all others contributing to the confusion of human phylogeny, has been the tendency for taxonomic individualization; as a result the species and genera of the fossil hominids have proliferated far beyond the limits

TABLE 1.4
The Chronology of Hominids

Epoch	Approximate time (years B.P.)*	Primate type
Pleistocene		
Upper Pleistocene	10 000–100 000	Homo sapiens sapiens
		H. sapiens neanderthalensis
Middle Pleistocene	100 000–1 000 000	Homo erectus
Lower Pleistocene	1 000 000–2 500 000	Australopithecines
Pliocene	2 500 000–15 000 000	Australopithecines
		Ramapithecus
Miocene	15 000 000–25 000 000	Dryopithecines
Oligocene	25 000 000–35 000 000	Propliopithecus
Eocene	35 000 000–55 000 000	Tarsiers and Lemurs
Palaeocene	55 000 000–65 000 000	Insectivore–prosimian

*B.P. – before present

of scientific validity (Simons, 1963). Historically, the publication of essays by Colbert (1949), Mayr (1951) and Simpson (1953) figure principally in the new awareness of the fundamentals of evolutionary systematics. Among others contributing to the developing comprehension of fossil man was Clark (1955), in his discussion of Morphological and Phylogenetic Problems of Taxonomy in

Relation to Hominid Evolution, in which he recognized the importance of the trait complex. Implicit in the concept of the total morphological pattern is that the characteristic occurrence of trait syndromes should form the basic unit of study; discrete traits are not important as analytical units. However, because of the fragmentary condition of many of the fossils the scope of the trait complex is of course limited, and in most cases it is confined to the size and morphology of the cranial bones and teeth.

The history of palaeoanthropology is indeed a cautionary tale, for with the new awareness of systematics and, through critical assessment, of those methods employed in the evaluation of fossils, the exposure of the Piltdown forgery was made (Weiner, Oakley and Clark, 1953). Hence an economy of hypothesis is advocated and all reliable taxonomic material and its phenetic affinities should, if possible, be classified under a single coherent scheme. Indeed Occam's principle applies as much to phyletic reasoning as it does to other scientific methodologies.

A complete review of the fossil record and interpretations obviously extends beyond the intended boundaries of this text; interested students could with advantage consult the book *Human Evolution* (Campbell, 1965) for a more comprehensive account. However, a brief summary of hominid evolution now follows whilst Table 1.4 summarizes a time scale for the evolution of the primates.

HOMINID EVOLUTION

Australopithecus Stage

With Dart's original publication in *Nature* in 1925 – wherein he described the well known Taung skull from Bechuanaland, to which he gave the generic name Australopithecus – the taxonomic status of this, and related fossils since found, has been the centre of polemic anthropological scholarship.

The name Australopithecus means literally southern ape, and simply refers to the fact that this fossil hominid lived in the southern hemisphere. Since the time of Dart's discovery numerous fossils have been found consisting of crania, cranio-facial fragments and various parts of the post-cranial skeleton; collectively these have given us a fairly clear conception of what Australopithecus looked like. It would appear that they had attained erect posture and were capable of bipedal locomotion. Though they had small brains (one-third of the modern average), the cranial capacity varying between 450–650 cm^3, the ratio of brain size to body size suggests that the Australopithecines were mentally better endowed than their pongid contemporaries. As a group they stood on the threshold of humanity, and were savannah scavengers and opportunist hunters – possessing a rudimentary culture termed osteodontokeratic (Wolberg, 1970); that is to say they used naturally shaped tools of such materials as bone, horn and antlers. As to which population gave rise to the next evolutionary stage, *Homo erectus* – this is not known.

19

Homo erectus Stage

The first *Homo erectus* find was made by the Dutch anatomist Eugene Dubois in 1891 beside the Solo River in Java. Dubois gave his discovery the species name *Pithecanthropus erectus.* Since Dubois' time the legitimacy of his find has been confirmed by the discovery in Java and Africa of equally old and even older fossils of the same grade. On average, the size of brain was double that of the Australopithecines, and two-thirds that of modern man.

Because of the unevenness of the available data, assembled from various parts of the Old World — but principally from Java, China and Africa — it is rather difficult to summarize the general characteristics of the group. However, some tentative generalizations can be made concerning: (1) wide geographical distribution; (2) estimates of date, placing the Pithecanthropines in a time level from one-half to one million years before present (B.P.) — namely, in the early Middle Pleistocene epoch; (3) the possession of a lower palaeolithic culture which differs in form depending on the geographical location; (4) the remains at Choukoutien in China, which suggest that the use of fire had been achieved. (However it would be incorrect to extrapolate this feature to other geographical members of the group since no such evidence exists outside China); (5) a marked reduction in molars in comparison with the Australopithecines, and falling within the modern range of variation. This sharp decrease in molar size has been correlated with the development of more effective hunting techniques.

Neanderthal Stage

In 1856, the first Neanderthal fossils were discovered in a limestone cave in the Neander valley, near Dusseldorf, Germany. Since this time numerous other finds have been made indicating that the group had a wide geographical distribution. In general the Neanderthals appeared in the Riss-Wurm interglacial age more than 100 000 years ago. The most significant change in the neander group which evolved from the *Homo erectus* grade, concerned the increase in cranial capacity; this was now fully 'modern'. It is evident therefore that some kind of threshold had been reached, in that any further increase in brain size conferred little or no survival advantages upon the possessors. The Neanderthal fossils are found associated with a Middle Palaeolithic culture.

There are several phylogenetic schemes, each purporting to represent the course of evolution leading up to the emergence of modern man. *Figure 1.1* summarizes representative schemes. The author is of the opinion however that the so-called spectrum theory is by far the most satisfactory in explaining the evidence. Originally the theory was devised to explain only those fossils from the erectus stage upwards — but this has been extended to include the Australopithecines. Essentially the theorists suggest that there would exist at any one time in different areas groups exhibiting tendencies toward a higher grade. Also there would be intermediaries manifesting a mosaic of the grades, as exemplified by the Mount Carmel finds in Israel. At this site there is a confusing range of variation in the skeletal material (*see* Clark, 1955). Hence it would appear that

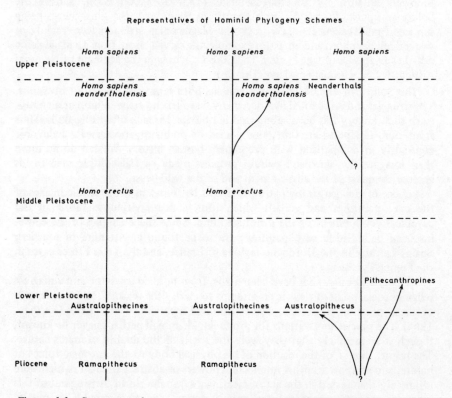

Figure 1.1. Summary of representative schemes of Hominid phylogeny (after Dr. Christopher Peebles, University of Windsor)

the Neanderthal population, like the earlier grades, were a highly polytypic species of Homo, widely distributed and deeply rooted in time; furthermore the evidence indicates that they had adapted to several environments, manufacturing a variety of implements from which we can conclude that they lived in different cultural contexts. Thus those populations exhibiting higher grade tendencies gave rise to forms of a higher level.

It is not known when or where the first men of the fully modern type — *Homo sapiens sapiens* — appeared, though generally it is considered that they made their first appearance toward the end of the final Wurm glaciation.

The Upper Palaeolithic period spanned some twenty millennia, from approximately 30,000–10,000 B.C. Among the first modern men were the Aurignacians, who were superseded by the Solutreans (20,000–15,000 B.C.) followed in turn by the Magdalenians (15,000–10,000 B.C.). During the Solutrean epoch, flint tools were beautifully fashioned, the finish never being surpassed in excellence. In contrast, the Magdalenian artefacts were mainly of worked reindeer horn and bone (for example, the carved bison, from La Magdeleine near Les Eyzies Dordogne – *circa* 15,000 B.C. – now in the Museum of National Antiquities, St.-Germain-en-Laye, France).

The Solutrean and Magdalenian epochs, both represent migratory invasions. The struggles and inter-tribal rivalries, if any, have left no trace in human memory, since such history has been submerged in oblivion for lack of writing skills. Thus from our standpoint in time, this period of prehistory possesses a halcyonic continuity in comparison with subsequent human history. We can do no more than imagine the slow and painful progress made by Palaeolithic man in his gradual conquest of the environment and of the wild beast.

Echoes of this gargantuan struggle are to be found in the indelible images of the cave paintings and graffiti which form a near-continuous fabric of the psychical development of the primitive mind. Some fine examples of these awe-inspiring palaeolithic cave paintings are to be found at Altamira in southern Spain, Lascaux in the Dordogne regions of France, and the Trois Fréres caves in the French Pyrenees.

Many analytical studies have been made, from both an aesthetic and a magico-religious social point of view, in an attempt to clarify the reasons for the existence of these paintings (Graziosi, 1960; Laming-Emperaire, 1962; Leroi-Gourhan, 1968). The precise motivations for producing them will perhaps never be known, though it seems likely that they were the result of the duality of man's nature. The result, that is, of the reaction of his physical body to the intrinsic forces of nature, superimposed on his intellectual powers of abstract thought, which were ultimately manifested in the act of creating a symbolic world. Arising out of this inherent need to connect form with function, and thus mould his mental imagery into dogma, magical rituals were formulated: these required the beast or figure to be drawn, so effecting their transformation into a symbolic vehicle for man's intellectual communion with those forces which he believed governed the destiny of his hunt and the sexual success of his tribal group.

The cave art of the Magdalenian period reached a high point of excellence and, in result, this is often referred to as the 'Classical' period. As the climate in Europe became increasingly mild, so the deer and other herbivore herds, upon which the Magdalenians depended, migrated further north, this resulting in the 'sudden' disappearance of the Upper Palaeolithic culture, and the end of an artistic tradition which had lasted 20 000 years.

In the Near East the Neolithic period followed (8,000–3,500 B.C.) in which man made the slow transition from hunting to husbandry. Up to this time man had led a hunting-gathering way of life, that involved a nomadic or semi-nomadic existence. The major transformation in this period was the domestication of plants and animals which permitted a more permanent way of life. With a sedentary pattern of living more attention could be given to enhancing the quality of life — hence a proliferation of the various arts and technologies, such as weaving, pottery, metallurgy, house building and other specialities. This revolutionary trend was accompanied by profound changes in Man's attitude toward himself and his environment. It also produced a social hierarchy and the peasant — for, prior to the development of civilization, there had been no peasants. The new agricultural civilizations in Egypt, Mesopotamia and the Indus valley exercised social control resulting in a shift to urbanization and the creation of political systems, socio-religious complexities and organized warfare — the most terrifying of all human occupations.

Clearly *Homo sapiens* is a social animal and, being so, required of himself a device by which he could communicate ideas to his fellow beings. Thus at some time in hominid evolution language appeared, the development of which paralleled human cultural activity in successive phases. Examination of fossil skulls reveals that the speech centres of the cortex were sufficiently developed in *Homo erectus* to suggest that they were capable of some form of speech. Increasing sophistication was inevitable, in that the spoken word was reinforced by the written word — the symbol of speech.

And whatsoever Adam called every living creature, that was the name thereof. And Adam gave names to all cattle and to the fowl of the air, and to every beast of the field (Genesis II: 19—20).

Man began, then, to classify his environment in an attempt to understand the experiences and knowledge gained by one generation and to pass it on to the next via the written word. McCluhan (1967) has written: 'Civilization is built on literacy, because literacy is a uniform processing of a culture by a visual sense extended in space and time by alphabet . . .'. Indeed it would appear that the very structure of language, the origins of which are in the organization of visual patterns, so emancipated the human brain that it functioned in a manner quite distinct from that of other animals. Thus the formalization of language became an instrument for thinking in much the same way as the ancient abacus gave rise to the computer, and with it the means to calculate in a completely new fashion. Furthermore, as implied earlier, urban man was the product of this new order. Lewis Mumford (1961) wrote in *The City in History:* 'The City is a special receptacle for storing and transmitting messages. The development of symbolic methods of storage immensely increased the capacity of the city as a container; it not merely held together a larger body of people and institutions than any other kind of community, but it maintained and transmitted a larger portion of their lives than individual human memories could transmit by word of mouth'.

To conclude this chapter, by scientific deduction — made on the grounds of recognizing homologous structures, comparative morphology and anatomy of animal and plant groups in relation to direct geological and palaeontological evidence – we are able by legitimate extrapolation to arrive at the conclusion that hominization of the biosphere was the inevitable culmination of some 4×10^9 years of random biological activity.

As to the future of *Homo sapiens* – the question must be left unanswered; but if viewed in context with the evolutionary processes it is not difficult to imagine man simply as a transient vehicle of energy. The teleological implication here is not one of theurgic forces directing the evolutionary pathway to the fulfilment of some higher design: on the contrary, it merely implies mechanistic purpose. That is the successful adaptation, or adjustment of a continuous self-organizational pattern or design to a new environment. The mechanistic point of view clearly dismisses the presence of special forces of nature as a determining feature of life. Vitalistic theories like those of Hans Driesch have endeavoured to explain the phenomenon of life by postulating such a directive force as being inherent in the living organism – though obviously the particular spatial inter-action of macromolecules, the collective configuration of which manifests life, possesses directive forces not seen in the inanimate world. To the mechanist this is simply seen as a matter of differential design. Recent events in molecular biology clearly reveal that no new force of nature as proposed by Driesch and other vitalists appears necessary to explain these life processes, and that both the animate and inanimate world are governed by the same physical laws. Clearly vitalism has been undermined, though it must be conceded that such views have not been entirely dismissed by all scientists.

The consequence of these unique qualities manifested by the human condition cannot be clearly foreseen – indeed, a truly objective interpretation is extremely difficult, if not impossible: this, since man as a biological entity forms an integral part of the biosphere which is inextricably entwined with the physical and inorganic totality of Nature. Thus it is the intention of this text to consider the basic unit of life in an attempt to understand some fundamental aspects of this special state of matter.

ACKNOWLEDGEMENTS

Thanks are due to Random House, Inc for permission to use excerpts from *The Immense Journey* (1957) by Loren Eisely; and to Dover Publications, Inc. for permission to quote from a translation by S. Margulis of *The Origin of Life* (1953) by A. I. Oparin.

BIBLIOGRAPHY AND REFERENCES

Bakker, R. T. (1971). 'Ecology of the Brontosaurs.' *Nature, Lond.*, **229**, 172.
Bernal, J. D. (1961). 'The Origin of Life on the Shores of the Ocean.' In *Oceanography*, Ed. by M. Sears. *Publs Am. Advmt. Sci.*, **67**, 95.
– (1967). *The Origin of Life*. London; Weidenfeld and Nicolson.

BIBLIOGRAPHY AND REFERENCES

Campbell, B. (1965). *Human Evolution: An Introduction to Man's Adaptation.* Chicago; Aldine Press.

Clark, W. E. Le Gros (1955). *The Fossil Evidence of Human Evolution.* Chicago; University of Chicago Press.

— (1965). *History of the Primates: An Introduction to the Study of Fossil Man.* London; Brit. Mus. (Nat. Hist.).

Colbert, E. H. (1949). 'Some Palaeontological Principles Significant in Human Evolution.' In *Early Man in the Far East: Studies in Physical Anthropology.* Ed. by W. W. Howells. *Am. Ass. Phys. Anthro.,* **1**, 103.

Dart, R. A. (1925). '*Australopithecus africanus:* the Man-Ape of South Africa.' *Nature, Lond.,* **115**, 195.

Dobzhansky, Th. (1962). *Mankind Evolving: The Evolution of the Human Species.* New Haven, Conn.; Yale University Press.

Du Toit, A. I. (1937). *Our Wandering Continents.* Edinburgh; Oliver and Boyd.

Eiseley, L. (1957). 'How Flowers Changed the World.' In *The Immense Journey,* pages 71—74. New York; Random House.

Fox, S. W. (1965). *The Origin of Prebiological Systems.* New York; Academic Press.

— and Harada, K. (1960). 'Thermal Copolymerisation of Amino Acids Common to Proteins.' *J. Am. chem. Soc.,* **82**, 37—45.

Garay, A. S. (1968). 'Origin and Role of Optical Isomery in Life.' *Nature, Lond.,* **219**, 338.

Glaessner, M. F. (1961). 'PreCambrian Animals.' *Scientific American,* **204**, 72.

Graziosi, P. (1960). *Palaeolithic Art.* London; Faber.

Haldane, J. B. S. (1929). *The Origin of Life.* London; Rationalist Press Association.

Kvenvolden, K. A., Peterson, E. and Pollock, G. E. (1969). 'Optical Configuration of Amino Acids in PreCambrian Fig Tree Chert.' *Nature, Lond.,* **222**, 1132.

Laming-Emperaire, A. (1962). *The Meaning of Palaeolithic Art.* Paris; Picard.

Leakey, L. S. B. (1968a). 'Upper Miocene Primates from Kenya.' *Nature, Lond.,* **218**, 527.

— (1968b). 'Bone Smashing by Late Miocene Hominidae.' *Nature, Lond.,* **218**, 528.

Leroi-Gourhan, A. (1968). 'Evolution of Palaeolithic Art.' *Scient. Am.,* **218**, No. 2.

Lissner, I. (1957). *The Living Past: The Great Civilizations of Mankind.* London; Jonathan Cape.

Mathews, C. N. and Moser, R. E. (1966). 'Prebiological Protein Synthesis.' *Proc. natn. Acad. Sci. U.S.A.,* **56**, 1087.

Mayr, E. (1951). 'Taxonomic Catagories in Fossil Hominids.' *Cold. Spr. Harb. Symp. Quant. Biol.,* **15**, 109.

McCluhan, M. (1967). *Understanding Media: The Extensions of Man.* London; Sphere.

Miller, S. L. (1953). 'A Production of Amino Acids Under Possible Primitive Earth Conditions.' *Science, N. Y.,* **117**, 528.

— and Urey, H. C. (1959). 'Organic Compounds' Synthesis in the Primitive Earth.' *Science, N. Y.,* **130**, 245.

25

Mivart, St. G. J. (1873). 'On Lepilemur and Cheirogaleus and the Zoological Rank of the Lemuroidea.' *Proc. Zool. Soc., Lond.,* **17**, 484.

Morris, D. (1967). *The Naked Ape.* London; Jonathan Cape.

Mumford, L. (1961). *The City in History.* London; Secker and Warburg.

Nagy, B. and Nagy, L. A. (1969). 'Early PreCambrian Onverwacht Microstructures.' *Nature, Lond.,* **223**, 1226.

Oparin, A. I. (1924). *Proiskhozdenie Zhizny.* Moscow; Moskovski Rabochii.

— (1938). Translation of Russian Ed., 1936: *Origin of Life.* London; Macmillan. (Republished in 1953. New York; Dover Publications.)

— (1957). *The Origin of Life on Earth,* 3rd ed. Edinburgh; Oliver and Boyd.

— (1965). 'The Pathways of the Primary Development of Metabolism and Artificial Modelling of this Development in Coacervate Drops.' In *The Origins of Prebiological Systems.* Ed. by S. W. Fox. New York; Academic Press.

Oro, J. and Kimball, A. P. (1961). 'Synthesis of Purines under Possible Primitive Earth Conditions I: Adenine from Hydrogen Cyanide.' *Archs Biochem. Biophys.,* **94**, 217.

— — (1962). 'Synthesis of Purines under Possible Primitive Earth Conditions II: Purine Intermediates from Hydrogen Cyanide.' *Archs Biochem. Biophys.,* **96**, 293.

Pilbream, D. (1968). 'The Earliest Hominids.' *Nature, Lond.,* **219**, 1335.

Pirie, N. W. (1937). 'The Meaninglessness of the Terms Life and Living.' In *Perspectives in Biochemistry.* Ed. by J. Needham and D. R. Green. London; Cambridge University Press.

Sarich, V. M. (1967). 'The Origin of the Hominids: An Immunological Approach.' In *Perspectives on Human Evolution.* Ed. by S. L. Washburn and P. C. Jay. New York; Holt, Rinehart and Winston.

Simons, E. L. (1963). 'Some Fallacies in the Study of Hominid Phylogeny.' *Science, N. Y.,* **141**, 879.

Simpson, C. G. (1953). *The Major Features of Evolution.* New York; Columbia University Press.

Smith, G. A. and Hallam, A. (1970). 'The Fit of the Southern Continents.' *Nature, Lond.,* **225**, 139.

Stubbs, P. (1965). 'The Oldest Rocks in the World.' *New Scient.,* **25**, 82.

Weiner, J. S., Oakley, K. P. and Clark, W. E. Le Gros (1953). 'Solution of the Piltdown Problem.' *Bull. Brit. Mus. Nat. Hist.,* **2**, 141.

Wolberg, D. L. (1970). 'The Hypothesized Osteodontokeratic Culture of the Australopithecines: A Look at the Evidence and the Opinions.' *Current Anthropology,* **11**, 28.

2–Some General Considerations of the Basic Unit of Life

(Emergence of the cell theory – Instrumentation and the cell – The cell theory)

EMERGENCE OF THE CELL THEORY

European man underwent a radical intellectual and material metamorphosis during the fourteenth and sixteenth centuries AD., which marked the establishment of Modern Western civilization. This transition was hastened by many independent events, of which probably the most important had been the development of the printing press—which brought about the rapid dissemination and exchange of ideas until then parochially located and isolated. Also contributing to this new culture had been Nicholas Copernicus' startling announcement in 1543 that the earth rotated daily about its own axis, and that the universe was not, as Ptolemy had suggested, a finite entity with a stationary earth at its centre. Thus the developing intellectual climate in Europe towards the end of the sixteenth century was one of liberalization – together with a reactionary feeling directed at certain facets of the teachings of the ancient world. Much of this had evolved from the 'catalytic' work of the Humanists, who had made available to contemporary scholars the writings of the ancient Greek philosophers and naturalists. The resultant stimulus was to encourage such people as the anatomists to question the conclusions of their dissections – for mediaeval and early Renaissance anatomists still rigidly conformed to the precepts of their ancient masters; in fact, to have done otherwise would have been considered an act of heresy. Yet in the same year as that of the Copernican revelation, a professor at the University of Padua in Italy – Andreas Vesalius* – in his work, *'On the Study of the Human Body'*, found the courage to question in a few instances the anatomical legacy of Galen's teachings. Galen of Pergamon had been personal

*An asterisk after a reference indicates inclusion of date purely to show historical development.

physician to Marcus Aurelius, and was the most revered of all the ancient anatomists – hence the magnitude of this criticism can be appreciated, But this – now seemingly modest – outburst was of fundamental importance, since it retrieved anatomy from the clutches of academic ritual. For more than a millennium anatomical vivisection had been merely reinforcing apparent truths, rather than acting as a means of expanding the frontiers of human knowledge. Hence the demeanour of Western man, with regard to his own condition and the universe in which he found himself, altered – such that his previous scholastic and monastic attitudes gave way to attitudes of scientific enquiry and criticism.

In 1628 William Harvey published his famous treatise, *On the Movement of the Heart and Blood in Animals,* in which he laid the foundations of the modern concept of anatomy, by showing that the rhythmic contractions of the heart pumped blood around a closed circuit within the body. The only omission Harvey made in his thesis was his failure to explain how it was that blood managed to traverse the apparent void where the arteries disappeared into the deep tissues and then mysteriously reappeared as veins. The link was found a generation later when Malphigi (1661) and Leeuwenhoek (1680) described the capillary bed with the aid of a microscope. Marcello Malphigi (1628–1694) is well known for his anatomical and physiological investigations and in fact he was the first to recognize the existence of the capillary bed, though this is attributed to Leeuwenhoek. In his publication *De Pulmonibus Observationes Anatomici* (1661) Malphigi makes definite reference to the capillaries as the connection between arteries and veins. This theory regarding anastomosis was not communicated to the Royal Society in London by Leeuwenhoek until December 28th, 1683 (Dobell, 1939; Committee, 1939).

The microscope functions as a vehicle by means of which the human nervous system is able to extend its visual experience of a particular environment which exists beyond the boundaries of its powers of resolution. Thus, following the advent of microscopical systems in the seventeenth century, Man was at last able to transcend a physiological barrier which had been imposed upon him since his creation.

The resolution of the naked eye is somewhere in the region of 1/10mm; hence, without the discovery and subsequent development of this optical extension – with the biological sciences restricted thus to the level of macroscopic observation – progress would eventually have been impossible. Indeed, the microbial world would have remained an unknown quantity, as would the sub-macroscopic reality of living tissue, and all that has evolved from man's subsequent visual exploration of the cellular landscape. Thus the discovery of the microscope can be seen to be as significant to human society as was the invention of the wheel.

Robert Hooke (1665), was among the first of microscopists to observe and record the landscape of that world existing beyond the resolution of the naked eye. In describing his investigations, '*On the Texture of Cork by means of Magnifying Lenses*', Hooke observed the cell spaces, separated from each other by cellulose walls: the appearance of these he likened to small box-like structures and he subsequently referred to them as cells, thus introducing the word into scientific vocabulary. Later in 1674, Antony Leeuwenhoek discovered by chance the microbial world whilst examining the emulsification in water of the scrapings he had made from a decayed tooth: '. . . . I must confess', he wrote, 'that the whole stuff seemed to me to be alive. . . . though they were so little withal, that 'twould take a thousand million of some of them to make up the bulk of a coarse sand-grain. . . .'. He then went on to describe what he called animalcules, which were probably micro-organisms and infusoria (Dobell, 1939).

For more than 150 years this remained the full extent of man's knowledge regarding the microscopical world and the nature of the cell.

The ancient Greeks, with regard to the macroscopical world, claimed that all animals and plants however complicated are constituted by a few elements which are repeated in each one of them. They thus recognized a homogeneity of design existing throughout the animal and plant kingdoms. As a result of microscopical development and many independent observations made at the beginning of the nineteenth century (Mirbel, 1802*; Oken, 1804*; Lamarck, 1809*; Dutrochet, 1824*; Turpin, 1826*), a cell theory slowly emerged, which recognized a reciprocal homogeneity of design permeating throughout the microscopical regions of living material.

In 1827, achromatic lenses for compound microscopes were developed by Lister, who together with Hodgkins studied the fine detail of blood cells and vessels. The observations of Robert Brown (of Brownian Movement fame) in 1831 established that the nucleus was fundamental to all living cells — with the exception of circulating human erythrocytes and bacteria (*see* common features of cells, page 35). Such observations culminated with the studies of Schleiden (1838)* and Schwann (1839)*, both of whom independently propounded their cell theories, thereby crystallizing contemporary ideas regarding the living cell. It was in fact Schleiden who established the cellular nature of plants; his work stimulated Schwann to take up the cellular study of human tissue which finally led to Schwann's statement: 'Cells are organisms, and entire animals and plants are aggregates of these organisms arranged according to definite laws'. Thus the image of the cell at this time was merely one of a protoplasmic mass containing a central nucleus, limited in space by a cell membrane (*see* basic unit, page 36).

Both Schleiden and Schwann believed that a formative fluid called the cytoblastema gave rise to a nucleus which in turn condensed into the cell cytoplasm. In other words they were perpetuating a belief in some kind of intracorporeal spontaneous generation. But in 1859 Virchow* disproved this theory when he demonstrated that all cells were derived from other pre-existing cells. At the same time Kolliker* showed that cells were able to develop from the fusion of two cells – that is, by the union of sex cells, namely the spermatozoon and the ovum. Similarly the work of Flemming in 1880*, who described the phenomenon of mitotic division, and that of Waldeyer in 1890*, who showed that chromosomes were equally divided between the daughter cells during mitosis, consolidated the cell theory and added weight to the great debate surrounding the concepts of Darwin and Pasteur – principal characters in what might be termed the drama entitled 'The Age of Scientific Materialism'. All this serves to indicate the slow growth of knowledge dependent on the exchange of ideas and spirit of scientific enquiry formulated during the Renaissance.

The development of achromatic lens systems extended man's visual experience into even less accessible regions which in turn motivated the development of new techniques to further recognition of the newly revealed cellular landscape. For example, workers such as Ehrlich (1881)* introduced the technique of vital staining using methylene blue, and Romanowsky (1891)* devised an aqueous polychromatic dye system capable of fixing and staining the cellular components of blood smears.

Thus by the first quarter of the twentieth century the gross morphology of the living cell had been described. The following years have seen the development of the electron microscope and x-ray diffraction techniques which have extended our experience to the point at which we are able to examine objects separated by the extraordinarily small distance of 5 Ångstrom units. But as yet these techniques cannot be used to examine living cells which are, after all, the ultimate reference point of all biological investigation.

INSTRUMENTATION AND THE CELL

Fundamentally the simple microscope consists of a convex lens (or a combination of lenses) which when placed between the eye and the object being observed will result in an enlargement of the object-image formed on the retina. The degree of enlargement is the magnification of the instrument.

The human eye has the ability to resolve or distinguish between closely juxtaposed points in a visual field – in fact the limit of resolution of the naked eye, at a distance of 10 inches, is somewhere in the region of 0·2mm. The lens of the eye focuses light reflected by an object, and then projects this image on to the retina, which is a matrix of specialized light-sensitive rod and cone cells. These cells contain a pigment called rhodopsin which becomes excited when light energy is absorbed, with, in result, the transformation of this energy into electrical energy, and the subsequent initiation of a nerve action potential which is relayed to the visual centres in the brain giving rise to a sensation of vision. The distance between these rods and cones is some 5 μm – hence the eye is unable to resolve detail on the retina image at a distance any closer than 5 μm. Thus an object viewed under the microscope is brought closer to the eye, such that the image appears to the retina to be larger, and so experiences an increase in definition not otherwise appreciated.

The magnification of a microscope is the produce of two separate magnifications – that of the objective and of the eye-piece respectively. These are dependent on several factors: (1) the magnifying power of the eye-piece; (2) the focal length of the objective; (3) the length of the optical tube.

The magnification of an objective is obtained thus:

$$\frac{\text{Size of image}}{\text{Size of object}}$$

$$= \frac{\text{Distance of image from objective}}{\text{Distance of object from objective}}$$

$$= \frac{\text{Mechanical tube length}}{\text{Focal length of objective}}$$

The total magnification of a microscope is thus:

$$\frac{\text{Tube length (mechanical or optical)}}{\text{Focal length of objective}} \times \text{Eye-piece magnification*}$$

The capacity of an optical system, such as a light microscope, to distinguish between points very close together in an object is dependent on the wavelength (λ) of the light employed, and the numerical aperture (NA) of the objective lens. The minimum resolvable distance between two points (x) is given by:

$$x = (0.61 \times \lambda)NA$$

The numerical aperture (NA) is the sine of half the angle (α) formed between the centre point of an object and the edge of the objective lens, when the object is in focus, multiplied by the refractive index (n) of the transmitting medium. Thus:

$$NA = n.\sin \alpha$$

Since the sine cannot exceed 1, and the refractive index of most optical material does not go beyond 1·6 – the maximal numerical aperture of lenses for use in air is 0·95, and of oil immersion lenses, 1·4.

The term definition refers to the ability of an objective to produce a clear and distinct outline of the object-image, a capacity dependent on the elimination of spherical and chromatic aberrations.

*The magnification of the eye-piece is generally engraved on the mount by the maker.

Because of the size of a living cell, and its inherent transparency to visible light it has been an extremely difficult structure to observe. Success in the observation of such structures has therefore been dependent on the development of optical systems capable of increasing both resolution and contrast (to counteract cell transparency). This inability of an ordinary optical system to overcome transparency, due to low light absorption, is caused largely by the cell's high water content which can be by-passed by simply employing selective dyes which will stain particular areas of the cell structure — thereby introducing contrast by differential light absorption. But this technique has several major disadvantages — namely that fixation, the subsequent processing of the tissue and staining all precipitate irreparable chemical and morphological changes in the cell. Hence, in order to study a living cell, alternative means of introducing contrast have to be used.

Phase-contrast Microscopy

Phase-contrast microscopy is a technique wichh exploits intracellular differences in refractive index and thickness. These subtle differences in the cellular medium are interpreted in a phase-contrast system as a variation in intensity of brightness — such that what originally appeared to be a featureless area in an ordinary bright-field microscope will become a defined contrast area. Fundamentally the structure of the phase-contrast microscope is very similar to that of the ordinary microscope the difference being that it possesses several refinements; namely, a special substage condenser and phase objectives. The condenser incorporates a rotating metal disc carrying a series of annular diaphragms which complement an appropriate objective — this being dependent on the numerical aperture. These annular diaphragms are discs of opaque glass. The phase objectives differ from ordinary objectives simply in possessing a phase plate which is a glass disc that has a circular etched trough of a particular depth — such that when a light beam passes through it, it will have a phase difference of one quarter of a wavelength compared with the rest of the plate area.

The phase-contrast system thus utilizes the fact that, although cell structures are highly transparent, they will nevertheless cause small phase changes in transmitted light. For if the refractive index of an intracellular component through which a travelling light beam passes is higher than that of the surrounding medium then the light beam will be diffracted. In principle then the direct light from the annular diaphragm situated below the condenser will pass only through the trough in the phase plate of the objective. Hence if an object is interposed, undeviated (direct) light will still pass through the annular ring but deviated (diffracted) light will follow a slightly different pathway through the thicker glass of the phase plate outside the trough. This results in a retardation of the light wave by one quarter of a wavelength. Consequently when these two light waves re-emerge (one direct and the other diffracted) they unite out of phase — the difference making itself apparent in the form of appreciable changes in the intensity of brightness. These changes in phase are seen to increase in direct proportion to the differential refractive indices of the intracellular components and their thicknesses in relation to the surrounding medium.

For more definitive details of this technique the reader is advised to consult a more specialized work (*see* Bibliography).

Phase-contrast microscopy is a most valuable technique, and is applied routinely in the observation of cell cultures.

Dark Ground Illumination

Another means of introducing contrast is by the use of dark ground illumination. It is a method which renders visible an object otherwise not seen with an ordinary light microscope system. This is achieved by illuminating an object with oblique light, utilizing a special condenser — such that only light which is refracted by those objects having a refractive index greater than that of the surrounding medium will enter the microscope tube. As a result the object will appear brightly illuminated against a dark background. This method is widely used in the field of microbiology.

The Electron Microscope

The electron microscope is the only instrument at present available permitting a direct study of the cellular ultrastructure. Fundamentally it consists of an electron source which is generally a tungsten filament. If excited by a current of high voltage a continuous stream of electrons is emitted from this filament and these can be directed and focused in the plane of the object by a magnetic coil which acts rather like the condenser in an ordinary light microscope. This high velocity electron beam is then deflected by another magnetic coil which functions as an objective lens, this giving rise to a magnified image of the object. A further electrostatic field then serves as an ocular lens which magnifies the objective image. The final image can then either be projected on to a fluorescent screen for direct visual examination or on to a photographic emulsion for permanent record.

The resolution of an ordinary light microscope was seen to be dependent (*see* page 31) on the wavelength of the light employed and the numerical aperture of the objective lens. De Broglie's equation in which he postulated that a stream of electrons must obey the laws of quantum mechanics states:

$$\lambda = h/mv$$

when λ is the wavelength of a matter wave, h is Planck's constant; m is mass and v is velocity. It can be seen that the wavelength of a stream of electrons is a function of the voltage acceleration to which the electrons are subjected, and that the resultant λ is much shorter than that for light. That is, λ = 0·05 Å units for electrons and 5 500 Å units for light. Consequently, since the resolving capacity of a microscope is limited by the wavelength of the light employed in forming the image, the shorter the wavelength, the finer the detail that can be resolved — as is the case with the electron microscope. The limit of resolution with an electron microscope at present is in the region of 10 Å units. The resolving capacity is such that the image from the objective can be so enlarged as to achieve a total magnification of × 20 000.

X-ray Diffraction

Basically, x-ray diffraction is a technique consisting of passing a beam of monochromatic x-rays through a material such as a crystal and recording the resultant diffraction pattern — if any — on a photographic plate.

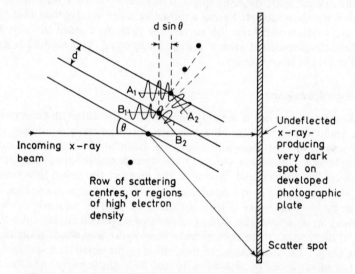

Figure 2.1. This figure illustrates the effect of an x-ray beam incident upon two parallel planes in a crystal lattice which are separated by a distance d. Using geometric principles, Bragg's Law ($n\lambda = 2d \sin \theta$) can be deduced and the distance (d) calculated

Consider a crystal structure as a three-dimensional lattice in which the constituent atoms are spaced at regular intervals along the three principal axes. If the crystal is then rotated in the path of an x-ray beam such that the planes are brought in turn to the correct angle to the incident beam, then at the centre of the photographic plate, on development, there will be a dark area due to undeflected x-rays — and a pattern of concentric spots corresponding to the scattering angles. The distance between these spots and the dark centre depends upon those spaces between the repeating units in the crystal or specimen which produced the diffraction.

The effect of an x-ray beam upon a series of scattering centres in a crystal lattice which are separated by distance d is shown in *Figure 2.1*. The incident rays A_1 and B_1 form an angle with these planes producing diffracted rays here designated A_2 and B_2. The dots represent scattering centres and θ the angle between the incident beam and the plane perpendicular to the row of scattering centres. Using Bragg's equation:

$$n\lambda = 2d \sin \theta$$

where $n = 1, 2, 3 \ldots \ldots$ etc, d can be deduced; thus, knowing the wavelength (λ) and the angle of incidence (θ) of a definite spot in a diffraction pattern the spacing producing the diffraction can be calculated.

The x-ray diffraction technique provides the molecular biologist with a means of studying the three-dimensional structure of biological molecules; — he can thus determine their precise position in a structure, and those distances separating them.

THE CELL THEORY

The single cell represents the simplest integrated organization in a living system which is capable not only of existing independently, but also of replacing its own substance in an appropriate environment.

All cells of both animals and plants exhibit basic common characteristics which are: (1) an ability to utilize extraneous energy, employing it to organize 'food-stuff' molecules obtained from the external environment to synthesize macromolecules specific to their particular somatic design; (2) an ability via cellular multiplication to perpetuate this synthesis information; (3) the means of channelling energy into different specific cellular activities, which serve to adjust and regulate the metabolism whereby the living cell in turn regulates its energy output to match its synthesis activity. That is, they are able to manifest a homoeostatic function.

With the development of the electron microscope it has been demonstrated that there exists a fundamental division among the protista based on the complexity of cellular organization:, namely, a division into eucaryotic and procaryotic cell types. In order therefore to discuss the differential characteristics of these two cell types it will be pertinent at this point to consider the hypothetical Ideal Cell, and – for the present purposes – merely regard all other cells occurring in nature, as a variation upon this theme. The Ideal Cell is depicted in Figure 2.2.

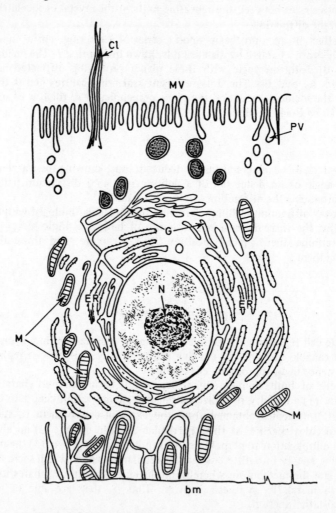

Figure 2.2. The ideal cell: cl = cilium; ER = endoplasmic reticulum – which is speckled, indicating ribosomes; G = Golgi region (note that it is smooth); M = mitochondria; MV = microvilli; bm = base membrane; N = nucleus; PV = pinocytic vesicle (from Spencer, 1970; reproduced by courtesy of Butterworths, London)

The Ideal Cell

The Nuclear Region

Generally the cell has a diameter of the order of 20 μm (2×10^5 Å units), and a volume of some 5 000 μm^3. Using a light microscope (and particularly with a phase-contrast system) the most easily identifiable structure within any cell is the nucleus: that has an approximate diameter of 5 μm and a volume of 40 μm^3. The interphase nucleus usually occupies a central position in the cell mass, but is capable of movement, and therefore may assume a variety of intracellular positions.

The nuclear region is highly organized, and often contains one or more distinct denser areas or structure called nucleoli. These areas are rich in RNA — in fact approximately 20 per cent of the total cellular RNA appears to be localized in these dense centres. Almost all the cellular DNA, on the other hand, is to be found in the nucleus; this is distributed throughout the nucleoplasm as chromatin. The nucleus is thus a cellular area limited by an envelope composed of a double membrane termed the perinuclear cisternae.

The outer wall of this membrane possesses ribosome particles, and it is for this reason that it is believed to form part of the endoplasmic reticulum. The perinuclear cisternae enclose a nuclear fluid, the nucleoplasm, which bathes the fabric of inheritance, this being a continuous network of DNA threads attached to a protein structure base, known collectively as chromatin. During mitosis this chromatin complex becomes organized into distinct structures called chromosomes. The number of chromosomes per somatic cell is constant (*see* Chapter 6).

Endoplasmic Reticulum

Arising from the perinuclear cisterna and continuous with the cell membrane is an anastomosis of canals known as the endoplasmic reticulum. It consists of lipoprotein membranes forming a compartment called the intracisternal space. Two types of endoplasmic reticulum have been distinguished, depending on the presence or otherwise of spherical ribonucleoprotein particles attached to the membrane surfaces: the former is termed rough; the latter, lined either with few particles or none, smooth. Endoplasmic reticulum appears to be largely concerned with the process of amino-acid assembly to form proteins and in other complex biosynthetic activity. This canal complex is also associated with the transportation of these products to their utilization or storage sites in other cell areas. Other functions appear to include synthesis control, ingestion of complex molecules and acting as a precursor for other membrane systems such as the Golgi region, For decades the Golgi region was believed to be an artefact, but it is now seen to be part of a common system arising out of the endoplasmic

37

reticulum. These Golgi complexes are more prevalent in those cells – such as the exocrine cells of the pancreas – which are associated with a secretory function. It is interesting to note that Golgi vacuoles can develop into secretory granules such as zymogen granules which contain and store proteins produced by the ribosomes of the rough endoplasmic reticulum (*see* Chapter 4). The true function of this region, however, is still shrouded in mystery.

The ribosomes which are intimately concerned with the protein synthesis activity of the rough-surfaced reticulum, have a sedimentation constant of 75S and 80S. They consist of two unequal sub-units with sedimentation coefficients of 35–40S and 55–60S respectively. Since they are difficult to crystallize, and cannot yet be studied in detail by x-ray diffraction techniques very little is known about their ultrastructure.

Mitochondria

The mitochondrion will vary in size and shape, but typically it is sausage-shaped. The average dimensions of the long and short axis are 1·5 and 0·5 μm respectively, with an average volume of some 0·8 μm^3. It is constructed of two membranes – the outer membrane enclosing the whole organelle, the inner membrane being invaginated to form internal partitions known as cristae mitochondriales which penetrate deeply into the intramitochondrial matrix. Linked to this inner membrane and the cristae are numerous small attachments consisting of a base plate, a stalk and head piece some 80–100 Å units in diameter; these have been called elementary particles (Green, 1964).

Essentially the mitochondrial unit figures as the site of intracellular energy production and transduction (*see* Chapter 5). Most of the enzymes of the citric acid cycle and the associated oxidative enzymes (for example the pyruvic dehydrogenase complex) and the β oxidation system for fatty acids appear to be localized on the outer membrane which also controls mitochondrial permeability. And the inner membrane and the elementary particles appear to be the site of those enzymes associated with the respiratory chain which is the final common pathway via which electron derivatives of biological fuels flow to oxygen. The inner and outer membrane form a characteristic protein–lipid–protein sandwich structure. The mitochondria appear to be unattached to any other structural cell component – and thus are capable of shifting their position within the cell, withthe purpose of initiating energy-yielding reactions.

Cell Membrane

The living cell is bounded by a membrane which is in a constant state of flux. This structure is of fundamental importance since it represents the only contact the total organism has with the external environment, and therefore plays an integral role in the organization of energy transfer and transportation systems (*see* Chapter 4). Present studies have shown that the cell membrane is composed of approximately 40 per cent lipid, 50 per cent protein and 10 per cent carbohydrate (Chapman, 1968). It is important to remember that the morphology of a cell membrane is purely an expression of interior activity, since — being intimately linked to a dynamic system — its structure will vary throughout the execution of its duties. The controversy regarding the lipoprotein assemblies in the membrane structure is outlined in Chapter 4.

Eucaryotic and Procaryotic Cells

The eucaryotic cell represents a more advanced cell system, and is the unit of structure in the higher protista, plants and animals. The eucaryotic cell possesses a nuclear membrane, multiple chromosomes within each nucleus and a mechanism via which equipartition of the products of chromosomal replication in the daughter nuclei is ensured. By contrast the procaryotic cell, which is the structural unit of the lower protista, does not possess a nuclear membrane; here, the nucleus consists simply of a single chromosome strand. Further to this, unlike the eucaryotic structure it also lacks other membrane-bounded organelles. The principal differences between the eucaryotic and procaryotic cells are summarized briefly in Table 2.1

TABLE 2.1

The Differential Characteristics of Eucaryotic and Procaryotic Cells

Characteristics	Higher protista: (most of the algae, fungi protozoa — the higher animals and plants)	Lower protista: (bacteria and the blue-green algae)
	Eucaryotic cells	Procaryotic cells
Mitotic division	+	—
Chromosome number	Always greater than one	Unknown
Nuclear membrane	Present	Absent
Mitochondria	Present	Absent
Chloroplasts	+ or —	—
Amoeboid movement	+ or —	—

These apparent differences seem to indicate that the lower protista are a distinct evolutionary group, representing vestiges of a stage in the evolution of the cell (*see* Table 1.1). The eucaryotic protista, on the other hand, are at a more advanced stage of evolution—their intracellular organization being not unlike that of the cells of the higher plants and animals. All this serves to indicate that subsequent evolutionary advances depended on cellular aggregation, specialization and differentiation, rather than on radical changes in the design of the cell structure.

The higher Protista which includes the Protozoa and the Protophyta can be divided into groups on the basis of their metabolic activity — that is, whether they are autotrophic or heterotrophic. Autotrophs are those organisms capable of synthesizing all their required food substances from simple inorganic molecules. Heterotrophs, on the other hand, are dependent on the products of other living cells. The Protozoa are heterotrophic and the Protophyta tend to be autotrophic. The particular distinguishing feature between the Protozoa and the Protophyta is that the latter are capable of photosynthesis (*see* page 42). Structurally the Protophyta have rigid exoskeletons, whereas the majority of the Protozoa have freedom of movement — that is, they are motile.

The Multicellular Organism

The multicellular organism is a stabilized organization consisting of cell aggregations which have specific inter-spatial functional relationships, in which the individual cell units have undergone specialized development — performing either structural, protective, metabolic or reproductive functions, such that their aggregate function is directed towards the maintenance of the status quo of the total organism. In fact one of the most intriguing and important features of the higher animal and plant cells is that the diversity of specialization occurring within the total somatic structure of a multicellular organism originates from a single multipotential cell.

Cellular Organization

According to the second law of thermodynamics, all systems (where a system is an isolated collection of matter) in the universe tend to move to a maximum state of randomness or entropy in an irreversible process, such that all energy is undergoing constant degradation; thus, the ultimate fate of such systems is to reach a state of complete disorder or entropic doom. A living system therefore appears clearly to achieve the opposite of what is predicted by the second law of thermodynamics — in as much as the cell is able to assemble simple molecules from the external environment via specific biosynthetic pathways into macro-molecules, thereby obviously decreasing entropy within the system. However this apparent ambiguity is resolved simply by regarding the living cell as an open system, which exists in a dynamic steady-state. Thus, the rate at which components are formed within the cells is immediately counteracted by an equal rate of degradation, thereby maintaining a state of equilibrium. The subsequent decrease in entropy due to synthesis activity within the cell is achieved, therefore, at the expense of increasing entropy in the external environment. For when cells absorb 'molecular-food' from the environment which possesses a relatively low entropy, they undergo degradation by oxidative reactions, producing simpler end-products which have a higher entropy value. These end-products are then discharged into the external environment thereby increasing entropy. Thus the living cell has developed a molecular homoeostatic mechanism whereby the steady state of its system is able to adapt to environmental changes, and entropy production is maintained at a minimum level.

The basis of cellular organization then is seen to be resident in a co-ordinated flow of information energy and matter, the features of which will now be elaborated.

On examining the co-ordination of those processes fundamental to a living system it will be seen that, with an increasing complexity of design, there is a reciprocating complexity of those mechanisms controlling somatic structure and function, and all these processes whether concerned with energy production or biosynthetic activity are predominantly under the control of enzymes (*see* page 44 and Chapter 8 . Hence it is important to examine the production of these enzymes since they establish the tempo of the metabolic pattern of the living cell.

Fundamentally then, the energy-transforming activity of a cell can be best seen as a flow of energy from 'food-molecules' to those endergonic processes vital for the function and survival of the living cell. Such a pattern of metabolic activity is dictated by information stored and perpetuated in the cell nucleus in the form of a specific pattern of purine–pyrimidine base pairs along the gene thread forming a sequence which is precisely replicated at each cell division (*see* Chapter 6). Furthermore the replication of this genetical information (the nucleotide and amino-acid sequence) is determined by further information, together with an available source of the energy required for the chemical linkages; all these are resident within the cell superstructure. The simplified flow diagram in *Figure 2.3* shows the basic relationships between the informational, chemical and thermodynamic aspects of a living system.

Photosynthesis

The essential feature of photosynthesis is radiant energy. The source of this energy is the sun, the surface temperature of which is somewhere in the region of $6 \times 10^3\,^\circ K$ – and a portion of the energy trapped within the atomic nucleus of hydrogen is converted into helium and electrons via nuclear fusion:

$$4_1H^1 \longrightarrow He^4 + 2_1e + h\nu$$

In this process a quantum of energy is released in the form of gamma radiation, which is represented in the above equation by $h\nu$ (where h is Planck's constant and v is the wavelength of the resultant gamma radiation). There then follow a series of complex nuclear reactions resulting in the emission of gamma radiation in the form of photons or quanta of light energy.

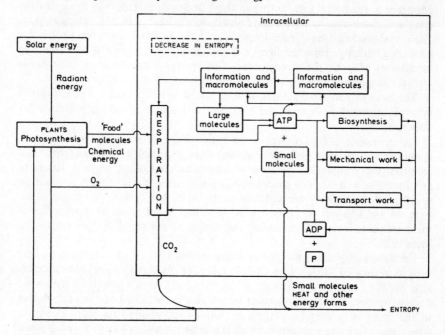

Figure 2.3. Simplified flow diagram indicating the basic relationship between the heterotrophic and autotrophic cells

Photosynthesis, then, is the ability of chlorophyll-containing cells to absorb this radiant solar energy and convert it into chemical energy: this is subsequently utilized for the reduction of atmospheric carbon dioxide and water to form carbohydrates.

The photosynthetic reaction may be simply written thus:

$$6CO_2 + 6H_2O + n.hv \longrightarrow C_6H_{12}O_6 + 6O_2$$

The glucose formed in this way by photosynthesis is then either converted to cell components — that is, cellulose, proteins and lipids — or oxidized to carbon dioxide and water to extract energy. These component-products of photosynthetic cells contain energy that can be utilized by the animal world, the members of which are not capable of photosynthesis.

To recapitulate, those cells capable of synthesizing complex molecular materials from simple environmental molecules are called autotrophs; on the other hand, the cells of the animal kingdom which are unable to perform this chemical manoeuvre, require the raw material products of the photosynthetic cells either as fuel or as material for the synthesis of cellular components — and for this reason they are called heterotrophs. Hence it will be apparent (*Figure 2.3*) that both heterotrophic and autotrophic cells are dependent upon the existence of a symbiotic cycle.

Heterotrophic Raw Material Requirements

Essentially the raw materials required to sustain normal heterotrophic cellular activity are lipids, proteins and carbohydrates. Cell activity can either take the form of energy expenditure (such as that required to transport molecules across a cell membrane); utilization, as in the storage of products which — following degradation at some other point in time — are capable of yielding energy; or alternatively the activity may be manifested as the synthesis of cell components. Work in a cell can take three distinct forms: (1) transportation and concentration; (2) mechanical; and (3) chemical. The latter — chemical work or biosynthesis — is performed by all living cells, not only during active growth periods, but also to maintain cell status. Hence the carbohydrates and lipids act as raw materials for the synthesis of other substances, and also as an energy source. This is true also of the amino acids which figure largely, too, in the formation of nucleotides — the sub-units of nucleic acids.

Following the integration of these three basic raw materials into the metabolic pathways of the cell a focal point is attained either directly or indirectly, at which stage pyruvic acid is formed. The breakdown of this substance is achieved through a complex sequence of oxidation reactions which yield energy and waste products (hydrogen and water). This area of metabolism is known as Krebs cycle, and takes place intracellularly within the mitochondria. Pyruvic acid and other acids are delivered to the mitochondrial sites where they are oxidized. At various stages in the cycle hydrogen is removed by the coenzymes NAD and NADP, and these enter the flavoprotein-cytochrome system which consists of a sequence of spatially related carrier substances along the shelves and inner membrane of the mitochondrion wall. As a consequence NAD and NADP transfer electrons down the enzyme chain, and for every pair which passes along the chain three adenosine triphosphate molecules are formed. Thus as a result of cell metabolism — the utilization of biological fuels — energy is acquired from oxidation reactions and made available to the system in the form of adenosine triphosphate (ATP), by the regeneration of adenosine diphosphate (ADP) and inorganic phosphate.

Informational Macromolecules

Fundamentally the nucleic acids are constructed from recurring mono-nucleotides in which the basic unit consists of either a purine or a pyrimidine base attached to a carbohydrate molecule (ribose or deoxyribose), which in turn is linked to a phosphate group. The nucleic acids are constructed using ATP as an energy source to facilitate specific sequential nucleotide arrangement dictated by information transmitted by the presence of the omnipotent DNA molecule.

The nucleic acids are molecules adapted for the transcription and storage of cellular information. Deoxyribose nucleic acid (DNA) is the master informational molecule in the cell, and is resident in the nucleus. Structurally the molecule consists of two helices entwined about a common axis, connected by hydrogen bonds between the pyrimidine and purine bases which are positioned at right-angles to the axis of the helices. These hydrogen bonds always form pairs between the bases adenine and thymine, or guanine and cytosine respectively. The result is a spatial structure of a right-handed helix, approximately 20 Å units in diameter, and of variable length, which — by virtue of a specific and sequential arrangement of these bases — forms the basis of the genetic code (*see* Chapter 6).

During mitosis, each daughter nucleus receives one of the complementary strands of the parent DNA molecule; synthesis of a complementary strand is effected by means of a process called replication whereby the genetic information inherent within the molecule is both duplicated and perpetuated. Information stored within these macromolecules is utilized: (1) to replicate its own image; and (2) to transmit information resulting in the assembly of specific proteins (*see* Protein Synthesis, Chapters 6 and 8 and *Figure 2.3*).

Forms of Regulation

Cells have evolved complex mechanisms (such as inhibition mechanisms) enabling the system to respond rapidly to changes in the environment thereby ensuring that proteins are synthesized according to demand. That is to say, those enzymes catalysing reactions of a particular metabolic sequence are capable of being turned on or off. Most enzymes are proteins which operate in relation to specific reactant complexes whereby they facilitate an increase of reaction rate.

The specificity of an enzyme appears to be dependent upon its structural configuration to which a reactant has an affinity — that is the activation site. This activation site very often has several foci along the molecule's length. It is conjectured that these sites, situated along the enzyme structure, position the reactant in such a manner as to be conducive to a rapid rearrangement of bonds, since in all chemical reactions enzyme activity depends on the number of collisions between the enzyme molecules and those of the substrate. Generally, the number of enzyme molecules present compared to those of the substrate is much smaller — hence there are other factors influencing the rate of collision, which will enhance the reaction rate. Some of these factors are: pH, temperature, and concentration of enzyme and substrate.

44

The hydrogen ion concentration will affect enzyme activity; this is not surprising really since enzymes are proteins — that is they behave as Zwitterions, and have isoelectric points. Each enzyme thus has an optimum pH at which activity is maximal. Similarly temperature is also an important factor. By increasing the temperature, an acceleration in activity will occur, but this enhancement will occur only to the point at which the heat produces a degradating effect on the molecule — whereupon activity rapidly decreases and is finally totally inhibited. In homothermal organisms the optimal temperature is $37^\circ C$. The temperature coefficient (Q_{10}) of a chemical reaction is about 2. That is, for every $10^\circ C$ rise in temperature, the rate of reaction will be approximately doubled. Naturally this is taking into account the inhibition point which for most enzymes is $56^\circ C$. Similarly the velocity of a reaction is seen to be very much dependent on two other important factors: (1) the number of enzyme molecules; and (2) the concentration of the substrate.

The relationship between the velocity V of an enzyme reaction, and the concentration of the substrate [S] is shown in *Figure 2.4.*

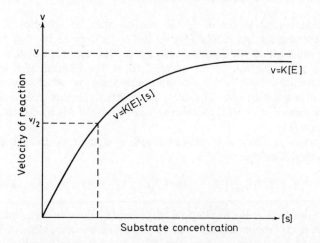

Figure 2.4. Depicts diagrammatically the relationship between the velocity (V) of an enzyme reaction, and the concentration of the substrate [S]

Cell metabolism requires the participation of hundreds of enzymes, all of which are carefully integrated and co-ordinated; the control of this is predominantly mediated via the enzymes themselves. One startling characteristic of enzyme kinetics is the phenomenal turnover of large amounts of substrate S by small amounts of enzyme E. This is explained by the fact that E combines with a reactant in the substrate S, forming an intermediary complex ES and is then regenerated at a subsequent stage together with the product P:

$$[E]+[S] \underset{k_2}{\overset{k_1}{\rightleftharpoons}} ES \xrightarrow{k_3} [E]+[P] \dots \text{ (1)}$$

From (1) the initial part of the reaction is:

$$k_1([Et] - [ES])[S] = (k_2 + k_3)[ES] \dots \dots \dots \text{(2)}$$

where [Et] is the sum of [E] and [ES]

subsequently, a rearrangement of (2) gives:

$$([Et] - [ES]) \cdot [S]/[ES] = (k_2 + k_3)/k_1 = K_m \dots \dots \dots \text{(3)}$$

where K_m is the Michaelis Menten constant.

In general for all values of S, the configuration of the plot of V versus [S] — a rectangular hyperbola (*Figure 2.4*) — is defined by two parameters V and K_s where V is the limiting velocity of the reaction which is shown in *Figure 2.4* as the plateau of the curve. That is, the formation of an enzyme substrate complex [ES] — representing the stage at which the enzyme is saturated with the substrate. The latter parameter K_s, equals that concentration of [S] for which v = V/2, this being known as the Michaelis constant K_m (*see* equation (3)).

Hence when an enzyme is saturated with a substrate — the velocity of the reaction is expressed thus:

$$v = k_3 [Et]/(K_m/[S] + 1) = V/(K_m/[S] + 1). \dots \dots \dots \text{(4)}$$

where v = velocity; V = the limiting velocity; and [S] the substrate concentration.

Since it is experimentally difficult to determine an exact value of V, the K_m can be found by making use of the following equation obtained by taking reciprocals of equation (4):

$$1/v = (K_m/V) \cdot (1/[S] + 1/v) \dots \dots \dots \text{(5)}$$

Thus the hyperbolic plot (*Figure 2.4*) can be recast in linear form by determining experimentally the velocities obtained with increasing substrate concentrations, and then plotting the reciprocals of V and S (*Figure 2.5* and equation (5)). Thus slope K_m/V cuts 1/v axis at 1/V and the 1/[S] axis at $-1/K_m$.

The above is one very simple example of enzyme control. Enzymes are also seen to be subject to other kinds of control, whereby the flux of materials through the myriad metabolic pathways of cellular activity is regulated with extreme precision. All of these mechanisms are conditioned by the status of the micro-cell environment, and the responses to these molecular 'messages' are conveyed directly or indirectly from one cell population to another by such means as the systemic circulatory system and the nervous systems (*see* Homoeostatic Function; Chapter 8 in Part V).

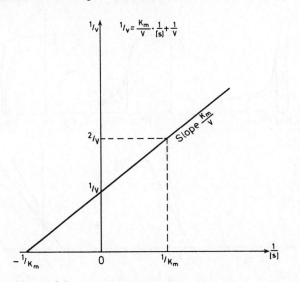

Figure 2.5. A reciprocal of the rate equation for an enzyme-catalysed reaction

One of these control mechanisms is product inhibition. Thus if products accumulate which inhibit an enzyme reaction then there occurs a deceleration of the enzyme reaction, and a corresponding rate of product accumulation such that equilibrium is restored.

Inhibitors may be either reversible or irreversible. An irreversible inhibitor acts until the enzyme can no longer perform, whereupon the reaction ceases prematurely. On the other hand reversible inhibitors may be either competitive or non-competitive. Competitive inhibitors affect the binding of the substrate to the enzyme by competing with the substrate for the enzyme-active site. The non-competitive inhibitors affect the reaction by binding to a site on the enzyme structure, other than the 'active' site; this, in addition to lowering enzyme activity, also slows the velocity of the reaction. In *Figure 2.6* the effects of competitive, non-competitive and mixed inhibition are illustrated.

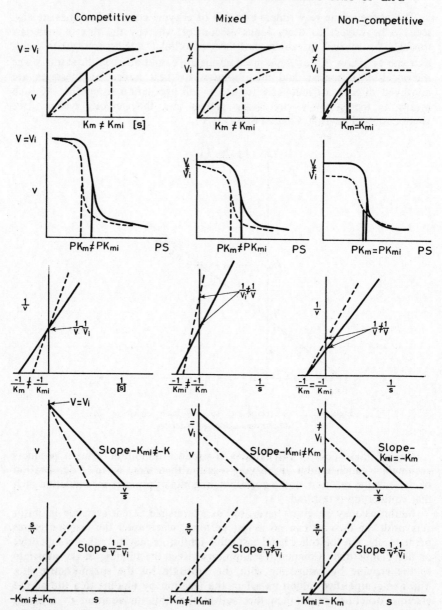

Figure 2.6. Illustrates the effect of competitive, mixed and non-competitive inhibition.
v = initial velocity; V = maximum velocity; i indicates the inhibitor. The solid line represents
the normal graph, and the dotted line the resultant effect of an inhibitor

48

Referring back to equation (1) we see that the formation of the product [P] depends upon the concentration of ES. Thus:

$$V_i = k_3 P; \quad \text{or} \quad V_i = k_3 . [E] \, [S] / K_m + [S]$$

Since [S] is very large compared to [E] essentially all the enzyme is present as [ES], and the reaction proceeds at a maximum velocity.

$$V_{max} = k_3 [E]$$

Hence
$$v = k_3 [E] / I + K_m / [S] (I + I/K_i)$$

where i is the concentration of inhibitor, and k is the dissociation constant of EI. That is:

$$E + I \xrightleftharpoons[k_i]{} EI$$

The above is the situation of competitive inhibition. Since $K_3(E)$ is not affected the V_{max} is unchanged. The corresponding equation for non-competitive inhibition is:

$$v = \left\{ k[E] / (I + I/k_i) \right\} \, / I + K_m / [S]$$

Here K_m remains unchanged whilst V_{max} is affected.

The importance of inhibitors is that they do not permit normal enzyme activity; this obviously will have a far-reaching effect on the human machine. In addition to this inhibitors are of great value in biochemical research.

Another type of control mechanism is feedback or retro-inhibition. Here the enzymes are directly inhibited by metabolic products far removed from the substrates of the reaction affected. Other forms of control are induction and repression. These will be discussed in Chapters 6 and 8.

BIBLIOGRAPHY AND REFERENCES

Bennett, A. H., Jupnik, Helen, Osterberg, H. and Richards, O. W. (1951). *Phase Microscopy: Principles and Applications.* New York: Wiley

Brachet, J. and Mirsky, A. E. (1961). *The Cell.* New York; Academic Press.

Brock, T. D. (Ed. and Trans.) (1961). *Milestones in Microbiology.* New York; Prentice-Hall.

Chapman, D. (Ed.) (1968). *Biological Membranes.* London; Academic Press.

Committee of Dutch Scientists (Eds.) (1939). *Collected Letters of Leeuwenhoek, Volume 2.* Amsterdam; Swets and Zeitlinger.

Cruickshank, R. (Ed.) (1965). *Medical Microbiology.* London; Livingstone.

Dobell, C. (1939). *Antony van Leeuwenhoek and his 'Little Animals'.* London; Staples Press. (Reprinted (1960) New York; Dover Publications.)

Driesch, H. (1921). *Philosophie der Organischen,* 2nd. ed. Leipzig; Engelmann.

Frankel, S. and Reitman, S. (Eds.) (1963). *Gradwohl's Clinical Laboratory Methods and Diagnosis,* 6th ed. St Louis; Mosby.

Green, D. E. (1964). 'The Mitochondrion.' *Scient. Am.,* **210**, 63.

Ham, A. W. (1965). *Histology,* 5th ed. Philadelphia, Pa.; Lippincott.

Harvey, W. (1628). *On the Movement of the Heart and Blood in Animals.*

Hooke, R. (1665). *On the Texture of Cork by means of Magnifying Lenses.*

Hughes, A. (1952). 'Some Historical Features in Cell Biology.' *Int. Rev. Cytol.,* 1; 1.

– (1959). *History of Cytology.* New York; Abelard-Schumann.

Malphigi, M. (1661). *De Pulmonibus Observationes Anatomici.*

Spencer, F. (1970). *Introduction to Human and Molecular Biology.* London; Butterworths.

Part II
The Molecular Basis
of Cell Structure

3– Fundamental Components of a Living Cell

(Amino acids – Characteristics and classification of proteins – Structural organization of proteins – Nucleic acids – Carbohydrates – Lipids and Steroids – Animal and plant pigments)

AMINO ACIDS – STRUCTURAL UNITS OF PROTEINS

The proteins are the omnipotent components of all living cells – performing vital functions in cellular architecture and in catalytic, metabolic and mechanical processes. They are polymers of amino acids, and approximately twenty different amino acids are yielded when proteins are completely hydrolysed. In Table 3.1 are listed those amino acids commonly figuring as structural units in proteins.

Essentially the amino-acid unit has an amino and a carboxyl function attached to the same α-carbon thus:

$$R.CH(NH_2).COOH$$

↑
The α-carbon

where R represents a specific side chain of variable length which will confer specific properties upon the resultant structure. By virtue of the simultaneous presence of both acidic carboxyl and basic amino groups the unit possesses amphoteric properties.

The amino acids can be classified according to the nature of the resident side chain: that is, according to whether they are (1) neutral; (2) basic; (3) acidic; (4) sulphur-containing; or (5) amides. For the most part proteins are constructed of linear, covalently bonded amino-acid chains. The behaviour of the amino acid – considering the inherent amphoteric property – will be directly controlled by the pH of the environment. The α-carboxyl group of the amino acid is characterized by a pKa value somewhere in the region of 2–3, whereas the

TABLE 3.1
The Principal Amino Acids which occur in Proteins

Name	Abbreviation	Nature of side chain	pI value	pK values COOH	NH₂
1. Glycine	Gly	Neutral	5·97	2·34	9·6
2. Alanine	Ala	Neutral	6·02	2·35	9·69
3. Valine	Val	Neutral	5·97	2·32	9·62
4. Leucine	Leu	Neutral	5·98	2·36	9·6
5. Isoleucine	Ileu	Neutral	6·02	2·36	9·68
6. Phenylalanine	Phe	Neutral	5·98	1·83	9·13
7. Tyrosine	Tyr	Neutral	5·65	2·20	9·11
8. Serine	Ser	Neutral	5·68	2·21	9·15
9. Threonine	Thr	Neutral	6·53	2·63	10·43
10. Tryptophane	Try	Neutral	5·88	2·38	9·39
11. Proline	Pro	Neutral	6·10	1·99	10·6
12. Arginine	Arg	Basic	10·6	2·17	9·04
13. Lysine	Lys	Basic	9·74	2·18	8·95
14. Histidine	His	Basic	7·58	1·82	9·17
15. Glutamic acid	Glu	Acidic	3·22	2·17	9·13
16. Aspartic acid	Asp	Acidic	2·87	2·09	3·86
17. Cystine	Cys	Sulphur-containing	5·06	1·65 7·85	2·26 (COOH) 9·85 (NH₂)
18. Methionine	Met	Sulphur-containing	5·75	2·23	9·21
19. Glutamine	Glm	Amides	5·65	2·17	9·13
20. Asparagine	Asg	Amides	5·41	2·02	8·5

amino group is nearer pKa 10. The pKa is that pH at which the group under consideration is in equilibrium between association and dissociation. Hence, in the pH range of 4—9 the amino acids exist as a dipolar ion — the carboxyl being dissociated and the amino group associated:

 Dissimilar polar groups attached to one carbon atom

Since most amino acids have a free acid and basic groups they can exist in aqueous solution as molecules with a net charge of zero, even though several groups may be ionized. These forms are known as Zwitterions. In such a case both positive and negative ions are present in the Zwitterion, such that the net charge on the molecule is zero.

Consider now the titration of the amino acid, glycine, from an acid to a basic solution. In the hypothetical titration the possible forms of glycine which could exist are:

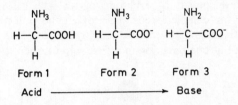

From this it will be seen that there are three forms of the molecule which can be present in the test environment during the change from acid to base. The middle form is the Zwitterion. From Table 3.1 it will be seen that there are two pK values for glycine, one for the -COOH and the other for the $-NH_2$, these values being 2·34 and 9·6 respectively. The pK of the group represents, therefore, the pH at which in that group one half of the molecules of that particular amino acid species are found as acid and the other half as a salt. Thus for glycine at a pH of 2·34 (*see above*) half will be found as form 1 and the other half as form 2, while at a pH of 9·6 equal amounts of 2 and 3 are present.

It has been observed that there is a definite pH value at which the sum of these positive and negative charges is zero. This pH is known as the isoelectric point, pI (*see* Table 3.1). At the isoelectric point proteins placed in an electric field do not migrate to either pole. Also at a low pH migration to the cathode occurs, whereas at a high pH migration is to the anode. This property is utilized in protein electrophoresis, which permits the separation of a protein mixture by virtue of differential component pI values. Similarly amino-acid mixtures can be separated using paper chromatography techniques. Fundamentally this technique requires a sample of a solution containing the amino-acid mixture to be applied to a filter paper – supported generally in a vertical position – which is then irrigated with a liquid organic solvent as n-butanol.

Paper chromatographic separations principally reflect differences in partition coefficients of the various amino acids between an organic and an aqueous phase. Irrigation is achieved by straight capillary action, or by a combination of the former and gravity. As the solvent passes the sample spot, it carries individual amino-acid components along with it at various speeds according to their solubility. As the solvent approaches the edge of the paper it is removed and dried. The individual amino acids of the mixture are then located by treatment with ninhydrin, when the developed areas appear as purple spots.

The individual amino acids separated from a mixture may be identified by comparison with pure standards run under the same conditions. The components are characterized by their Rf values — defined thus:

$$Rf = \frac{\text{Distance travelled by component}}{\text{Distance travelled by the solvent front}}$$

In practice development of a chromatogram with a single solvent system is not adequate for complete separation of all the amino acids. Hence two-dimensional chromatography can be employed. Here the chromatogram is initially developed in one direction in a first solvent; then it is rotated 90° and

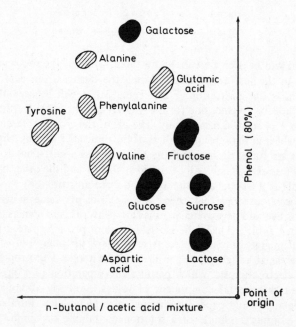

Figure 3.1. Illustration of a two-dimensional chromatogram
of amino acids

re-chromatogrammed in a second solvent system. Utilization of this technique affords a higher resolution of many components having closely related Rf values for a single solvent system. A two-dimensional chromatogram of amino acids is illustrated in *Figure 3.1*. Table 3.2 indicates those chemical reactions employed for the detection and semi-quantification of amino acids.

Finally before leaving this brief review of amino acids it is pertinent to mention that all amino acids obtained from proteins are optically active since the α-carbon of each is asymmetric — with the single exception of glycine. This absolute configuration of the amino acids is related to that of L-glyceraldehyde, and without exception it has been found that proteins digested in such a way that racemization does not occur are of this L-configuration.

TABLE 3.2
Reactions for Amino Acid Detection and Semi-quantification

Amino acid	Name of reaction	Reactant	Colour of reaction
Tyrosine	Folin-Ciocalteau	Phosphomolybdic-tungstic acid	Blue
Histidine or Tyrosine	Pauly	Alkaline sulphanilic acid	Red
Arginine	Sakaguchi	α-naphthol and sodium hypochlorite	Red
Cysteine	Nitroprusside	Sodium nitroprusside in dilute ammonia	Red
Tryptophane	Ehrlich	p-dimethylamino-benzaldehyde in conc. hydrochloric acid	Blue
Tryptophane or Phenylalanine	Xanthoproteic	Conc. nitric acid	Yellow
Tyrosine	Millon	Mercuric nitrate in nitric acid	Red

CHARACTERISTICS AND CLASSIFICATION OF PROTEINS

The condensation of amino acids to form protein molecules occurs in such a manner that the acidic group of one amino acid combines with an adjacent basic group — with a simultaneous loss of water; thus the resultant molecule possesses a basic and an acidic terminal unit (*see* the peptide bond, page 61). The general characteristics of the proteins are as follows.

(1) By virtue of their structure they retain the amphoteric properties exhibited by the amino acids — since an acidic group is always at one end, and a basic at the other. Similarly, owing to the possession of a facility for internal interchange of hydrogen ions at a certain pH, the molecule will carry no net charge — this is the isoelectric point (pI), and at this pH they are referred to as Zwitterions. At their pI values proteins will manifest:

(a)   solubility – the solubility of most proteins is a sensitive function of pH and the concentration of the organic solvent employed. Hence there is a great tendency for proteins to precipitate or coagulate;

(b)   stability as an emulsoid colloid;

(c)   minimal osmotic pressure and conductivity.

(2)   In addition to the amino-acid composition, the molecular weight is a basic piece of information employed to classify proteins. Accurate determinations of the molecular weights of most proteins have not yet been performed. Table 3.3 gives a short list of molecular weights of different proteins, these being determined from the colligative properties of the matter or from the rate of sedimentation. Among the colligative properties of matter only osmotic pressure is of real utility. It has been shown that those laws governing the mechanics of osmosis are analogous to the gas laws:

$$\Pi V = nRT \ldots \ldots \ldots \ldots (1)$$

Where Π is the osmotic pressure, V is the volume in litres of the solution, n is the number of moles of solute, T is the absolute temperature and R is the gas constant.

Hence from (1) Π can be related to the concentration (m) by the van't Hoff law:

$$\Pi/m = RT$$

Thus, where M is the molecular weight, and c is the concentration in grammes per cm^3:

$$\Pi/RTc = I/M \ldots \ldots \ldots \ldots (2)$$

(3)   Proteins also exhibit a high degree of specificity – as exemplified by the enzymes which are specific not only to the substrate they affect but also to the organism elaborating them.

TABLE 3.3
Some Representative Molecular Weights of Proteins

Trypsin	24 000
Pepsin	35 000
Insulin (tetramer)	47 000
Human haemoglobin	65 000
Human albumin	69 000
Human gamma globulin	156 000
Collagen	280 000
Pig thyroglobulin	650 000
Tobacco mosaic virus (TMV)	40 000 000

Purification of Proteins

Salt Precipitation

This technique is based on the principle that a given protein will be soluble to different degrees in different concentrations of a salt.

Isoelectric Precipitation

Here advantage is taken of the fact that as a molecule becomes less charged it will also become less soluble. Hence the solubility of a protein is least when the pH is at its isoelectric point — that is, that pH at which the net charge on the protein is zero.

Chromatography

In chromatography not only the charge on the protein but also the pH of the medium is taken into account. There are a variety of techniques which exploit this idea.

Column (ion-exchange) chromatography. — Here a column is set up containing ion exchange resins which have a particular electric character depending on the pH of the medium which is being employed. The solution containing the protein to be purified is passed through the column. The protein by virtue of its electrical nature will bind some of the groups to the resin. The protein can then be recovered by the passage through the column of solutions which have different pH values, since different proteins will be eluted out at different pH levels.

Gel permeation. — Here a column consisting of a highly polymerized carbohydrate matrix is used. This matrix serves as a sieve through which only those molecules within a certain size limit are allowed to pass. This is often used as a primary technique in purification.

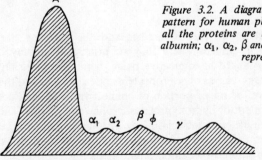

Figure 3.2. A diagram illustrating the electrophoretic pattern for human plasma measured at pH 8·6, where all the proteins are negatively charged. A: represents albumin; α_1, α_2, β and γ represent the globulins; and ϕ represents fibrinogen

Electrophoresis

Protein mixtures, as already implied, can be analysed by electrophoresis, which is the movement of charged particles of the protein species under the influence of an electrical field. An example of the type of data obtained – that is the electrophoretic pattern of a protein mixture contained in human plasma – is shown in *Figure 3.2*.

Crystallization

This technique is based on the assumption that crystals are pure products – though it is now known that this concept is not an entirely accurate one. This method of purification is often used in the final preparation of an already pure protein.

The proteins are divided into two general classes: (1) the simple proteins, which are defined as those proteins yielding only α-amino acids and their derivatives on complete hydrolysis; and (2) the conjugated proteins which are defined as those proteins containing one or more prosthetic groups.

The Simple Proteins

Protamines. – These are the simplest proteins, having a molecular weight somewhere in the region of 5 000 – 70 to 80 per cent of the total amino-acid constituent being arginine. These simple proteins are to be found in the mature spermatozoa of fish.

Histones. – The histones appear to occur in combination with the nucleic acids of somatic cells of many organisms (the nucleohistones). They consist of a somewhat greater variety of amino acids than do the protamines.

Prolamines. – These simple proteins are found only in plants, and are insoluble in water.

Glutelins. – As with prolamines these proteins are found only in plant material.

Scleroproteins. – In contrast to the prolamines and the glutelins, the scleroproteins are found only in animals where they are principally employed for architectural purposes – localized in connective tissue, bone, hair and skin, for example. There are two major classes of scleroproteins, the collagens and the keratins. The former constitute the major portion of connective tissue, and are characterized by their high content of hydroxyprolines, proline and glycine. The collagens are converted to soluble protein, gelatin, by prolonged boiling. The keratins on the other hand form the major constituents of epidermal structures, and possess varying amounts of sulphur-containing amino acids.

Albumins. – Unlike the proteins already referred to, albumins are characterized by their solubility in water and dilute aqueous salt solutions. They are also found widely distributed throughout the animal and plant kingdoms.

Globulins. – The globulins also exhibit a reasonable degree of solubility but this is limited to aqueous salt solutions. They form an important class of protein and, like the albumins with which they are generally associated, are widely disseminated in the animal and plant kingdoms.

Conjugated Proteins

Nucleoproteins. – The prosthetic group of the nucleoproteins is nucleic acid, which is allied to the protamine and histone classes.

Lipoproteins. – The lipid moiety of this group is variable, and the nature of the linkages between the lipids and proteins is still uncertain.

Mucoproteins. – This class of proteins has carbohydrates as its prosthetic groups, and again the nature and quantity of this component is variable; they are widely distributed.

Chromoproteins. – The prosthetic group of the chromoproteins is a pigment. The most important of these by far, at least as far as this text is concerned, are the respiratory haemoglobins, and the flavo-cytochrome pigments which are the principal agents in the oxidation system of the cell.

STRUCTURAL ORGANIZATION OF PROTEINS

As previously stressed the proteins are polymers of amino acids and possess high molecular weights. The structural features of these polymers are determined by the nature of the linkages which unite the monomeric units. The amino acids are assembled in fact by the formation of peptide bonds – linked through the -COOH and -NH$_2$ groups, with the elimination of water – one molecule per linkage (C* denotes the position of the α-carbon):

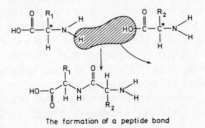

The formation of a peptide bond

61

The peptide bond thus establishes a backbone structure which is repetitive, the sequence of the residues within the chain conferring specificity upon the total structure. The polypeptide skeleton is seen, then, to consist of a sequence of nitrogen and α-carbon atoms from which the side chains and hydrogen atoms radiate. In *Figure 3.3* are shown dimensions and configuration of a fully extended polypeptide chain (Pauling, 1951), in which the structure consists of alternating

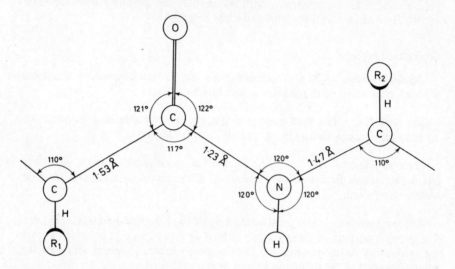

Figure 3.3. Bond angles and lengths of an extended polypeptide chain. [Ångstrom unit (Å) = 1/10 mμ; 1 mμ = 1/1000 μm; 1 μm = 1/1000 mm]

amide groups and α-carbon atoms of the amino acid. The two significant properties of the peptide bond from the standpoint of protein structure are: (1) the coplanarity of the peptide group; and (2) the ability of the amide groups to rotate with reasonable ease about the α-carbon.

The structures evolving from the formation of peptide bonds are called peptides, and the individual amino acids of the peptides are termed residues. Two amino acids linked together are called dipeptides. It will be noticed that the dipeptides still possess available terminal acidic and basic groups capable of producing tripeptides, tetrapeptides and so forth. Those aggregates consisting of up to 50 residues and which, as a consequence, have a molecular weight of up to 5 000 are generally referred to as polypeptides. Proteins, on the other hand, are structures containing from 50–5 000 or more amino-acid residues. The structure also may contain a prosthetic group.

When considering proteins it is convenient to visualize four levels of structural organization: primary, secondary, tertiary and quaternary. The distinction between these is made in terms of intermolecular interaction – such as hydrogen bonds, van der Waals forces, and salt linkages necessary for the maintenance of their structure.

Primary Structure

The primary structure of a protein is simply a linear polypeptide chain consisting of a known amino-acid residue sequence. The first successful description of a polypeptide residue sequence was only recently achieved, when

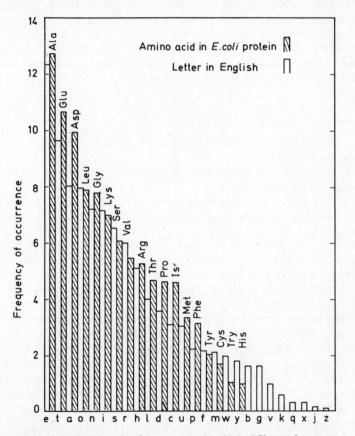

Figure 3.4. Showing the frequency with which different letters are used in written English and, by analogy, the frequency of occurrence of different residues in the amino-acid sequence of polypeptides (after Morowitz, 1958; reproduced by courtesy of the author and Pergamon Press Ltd.)

63

the complete primary structure of the polypeptide hormone insulin (molecular weight 6 000) was elucidated by Sanger (1952).

It would appear that for a given protein the residue sequence is the prime factor defining structural and intermolecular interaction — for as already seen a protein may contain from 50–5 000 or more amino acids assembled in a characteristic pattern peculiar to that particular protein from a pool of twenty different amino acids and positioned on the peptide backbone. The number of different permutations possible is almost infinite. In fact an analogy has been drawn illustrating how from twenty different amino acids it is possible to construct an infinite variety of sequences. Morowitz (1958) comparing the amino acids to letters occurring in a sentence showed that it was possible to construct innumerable sentences from the basic 26 letters of the English alphabet; he goes on to liken the frequency with which letters from the alphabet are used to the frequency with which the different amino-acid residues are utilized in the polypeptide structure (*Figure 3.4*).

Determination of Amino-acid Sequences

The way in which amino-acid sequences are determined generally follows the plan outlined below:

(1)   The molecular weight and the amino-acid component for the protein in question is initially determined.

(2)   The number of independent polypeptide chains is then determined.

(3)   Cleavage of interpeptide bonds and the separation of the individual chains is carried out.

(4)   Specific cleavage of the chain, and subsequent sequence of the residues of these individual cleavage products is then determined. When all the sequences of these degradation products have been determined then, by careful study, the complete sequence of the polypeptide chain can be established.

The primary structure of insulin which is the hormone concerned with the regulation of the rate of carbohydrate metabolism, was found to contain 51 amino-acid residues arranged in two chains (*Figure 3.5*). The chain terminating with glycine is called the A chain, and contains 21 residues in all, whereas chain B terminates in phenylalanine, and contains 30 amino acids. Following the elucidation of insulin, came the work of Hirs, Stein and Moore (1960), on ribonuclease at the Rockefeller Institute in New York. Ribonuclease is a single chain enzyme composed of 124 amino acids, with four interchain disulphide bridges (*Figure 3.6*).

Thus it has been deduced that there exists for protein species a specific residue sequence. For example (*Figure 3.5*), Sanger found that in mammalian insulins there occurred an amino-acid variation in either the A chain, between residues 8 and 10, or at the carboxyl terminus of the B chain. Similarly the

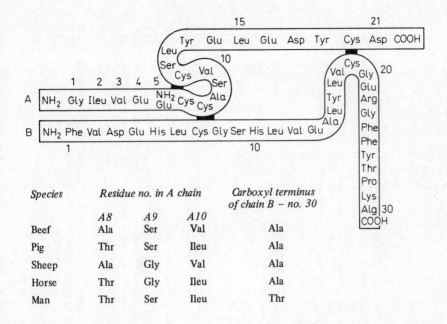

Species	Residue no. in A chain			Carboxyl terminus of chain B – no. 30
	A8	A9	A10	
Beef	Ala	Ser	Val	Ala
Pig	Thr	Ser	Ileu	Ala
Sheep	Ala	Gly	Val	Ala
Horse	Thr	Gly	Ileu	Ala
Man	Thr	Ser	Ileu	Thr

Figure 3.5. (Above) The primary structure of beef insulin: note variations in chains A and B – A having 21 residues and B, 30. (Below) Key indicating species variation in chains A and B

arrangement of the disulphide bridges (where they occur) is unique to a particular protein. Again, it is found that only amino acids of the L-configuration occur in proteins.

Secondary Structure

It was seen that the polypeptide backbone in the primary protein structure is a long flexible chain the adjacent molecules of which are held together at short range by van der Waals forces and that in particular at greater distances the molecules are bridged by disulphide bonds (S-S) between cysteine groups – as seen in the insulin molecule (*Figure 3.5*). Furthermore, these disulphide bridges are seen to be important in conferring rigidity upon the single chain structure.

The term 'secondary structure' refers to the regular configuration of the polypeptide backbone, which is due to the formation of hydrogen bonds between the compounds of the peptide linkage itself. The hydrogen bond arises between

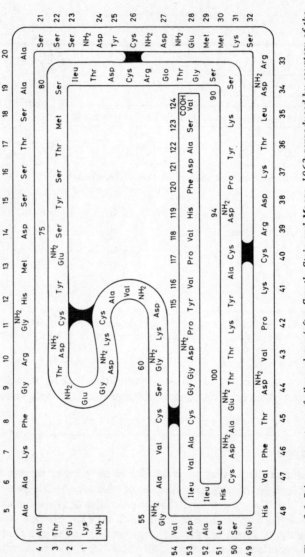

Figure 3.6. Primary structure of ribonuclease (after Smyth, Stein and Moore, 1963; reproduced by courtesy of the authors and the Journal of Biological Chemistry, Baltimore)

a covalently bound hydrogen atom with some positive charge, and a negatively charged covalently bound acceptor atom. For, electronegative atoms of elements such as oxygen and nitrogen have an electro-affinity for protons, which – by virtue of their size – permit attracted molecules to within a very short distance of themselves, thereby establishing a link known as a hydrogen bond. This sort of bond is more specific than that established by van der Waals forces, since the former demands the existence of a molecule possessing complementary donor hydrogen and acceptor groups – whereas the latter merely arises from a non-specific attractive force manifested when two atoms are in close apposition to one another.

In most amino-acid polymers, the side chains are very large, and as a consequence would seemingly distort such an arrangement as that proposed in the primary structure. Hence an alternative three-dimensional character is needed to accommodate them such that: (1) the peptide group be planar; (2) every carbonyl oxygen and amide nitrogen be involved in hydrogen bond formation; (3) these criteria are repeated, and are the same for every residue. Such a structure – meeting these specifications and so ensuring maximum stability – was found and termed the α helix (Pauling, Corey and Branson, 1951). Hence in the secondary structure of proteins the backbone of the molecule is twisted to form a right-handed helix – such that the -NH groups of the peptide bond on the one turn can form hydrogen bonds with an adjacent -CO group thereby giving the total configuration molecular stability. Examination of a molecular model of a peptide chain reveals that the amide groups still lie in a common plane, that rotation occurs about the α-carbon, and that the side chains (R) radiate from the helix superstructure. Proteins which contain α-helical structures are either globular or fibrous proteins.

The fibrous proteins are often arranged in an orderly fashion easily identified by x-ray diffraction techniques, and are classified into the collagen group, the fibrinogen group, myosin, α-keratin, β-keratin, fibrosin and silk.

The Tertiary and Quaternary Structures

With the successful analyses of the x-ray diffraction patterns of myoglobin, the oxygen-carrying protein of mammalian muscle (Kendrew *et al.*, 1958), it was shown that the component polypeptide chains were folded back on themselves within a globular protein resulting in a complex three-dimensional sausage-shaped form. The 'pathway' of this resident peptide chain is known as the tertiary structure – this term referring to a higher level of structural organization in which the polypeptide chains have undergone a precise folding operation yielding a supercoiled form. Apart from myoglobin the only other protein structure of comparable complexity to be studied successfully has been haemoglobin – but it is supposed that the other proteins will be found to form these supercoiled complexes, the precise pattern being dictated by their specific residue sequences.

Myoglobin has a molecular weight of 17 500, and consists of eight helical segments which for the most part occur in the linear areas of the structure, these being interrupted by the coiled regions. The whole structure is stabilized by hydrogen bonds and other cross-linkages. The single polypeptide chain is composed of 153 amino-acid residues and one prosthetic group, a haem group containing iron.

The haemoglobin molecule — the oxygen-transporting pigment of mammalian blood — though closely resembling myoglobin in structure, on the other hand represents a further degree of organization in possessing a quaternary structure.
In fact the haemoglobin molecule has a molecular weight of 65 000 which is four times that of the myoglobin molecule — and each molecule consists of four peptide chains which are identical pairs, designated α and β respectively, and are associated with the four resident haem groups. The α chain contains 141 amino-acid residues, compared with 153 in myoglobin. The β chain contains 146. Over most of the length of the sub-units, the general conformation is similar to that of myoglobin. It has been shown by x-ray diffraction studies (Perutz *et al.*, 1960; 1968) that the overall shape of the oxygenated molecule is spherical with a length of 64 Å, a width of 55 Å, and a height of 50 Å. The points of contact between the unlike units of the α and β chains are chiefly non-polar. Each haem group is attached to a molecular sub-unit pair, each of these being formed by a complex convolution of a single polypeptide chain. Simply, then, the total molecule may be represented thus:

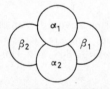

A molecular model of haemoglobin is seen in *Figure 3.7* — the white part of the model representing the α_1 chain and the black portion the β_2 chain. This photograph of the model clearly demonstrates how closely these chains fit together in an approximately tetrahedral arrangement in the quaternary structure of the molecule; it indicates also those segments involved in the contact of α_1 and β_2 chains. In fact the most extensive points of contact occur between α_1 and β_1 involving about 110 atoms, that is 34 contributing residues, coming within a distance of 4 Å of each other. The α_1 and β_2 contacts on the other hand only involve some 19 residues, contributing about 80 atoms.

The haem portion of the molecule consists of a tetrahedral arrangement of four pyrrole rings centred about a nucleus of iron. These units are bridged by

68

Figure 3.7. Photograph of model depicting the association of two chains of α-haemoglobin and two of β-haemoglobin to form the complete molecule of haemoglobin (reproduced by courtesy of Dr. M. F. Perutz)

four methine (-CH-) bridges; the nucleus being in the ferrous form is attached to the nitrogen atoms of the pyrrole rings. Also the iron is linked further to the nitrogen atoms of the iminazole group associated with the polypeptide (globulin) complex. Each single unit has a molecular weight of approximately 17 000.

Oxygenation of the molecule results in a marked alteration in structure, but this process is reversible. The ability of ferrous iron in haemoglobin to combine reversibly with molecular oxygen is apparently due to its non-polar surroundings. In fact though the total molecule does not appear to be affected by replacement of surface amino-acid residues, the reverse is noticed when small alterations are made to the non-polar contact regions of the α and β sub-units — particularly those in the proximity of the haem moiety. Structural differences in the tetrahaemic haemoglobin molecule found in man have enabled two categories of haemoglobins to be identified — the physiological and the pathological (see Chapter 6).

Those forces stabilizing tertiary and quaternary configurations such as myoglobin and haemoglobin are obviously multifactorial, but primarily the three-dimensional configuration represents the most favourable arrangement of the polypeptide chain, thermodynamically (Klotz, 1957). The reliance on the S-S bond in the sub-unit aggregations is seen not to obtain in the structures of a higher order such as haemoglobin, but instead the stability of the configuration appears to be dependent on the hydrophobic (or polar) nature of the hydrocarbon side chains, and the electrostatic forces set up between charged groups. For a discussion on the thermodynamics of such bond functions the reader is advised to consult the papers of Kauzmann (1959) and Klotz (1960), and specific biochemical texts (see Bibliography), since much of this is still speculative. To conclude, the spatial configuration of the higher order of proteins is to a large extent predetermined by the amino-acid sequence in the primary structure and by the bonds thereby established among the resident residues.

NUCLEIC ACIDS

The nucleic acids, that is deoxyribonucleic acid (DNA) and ribonucleic acid (RNA) are compounds of fundamental biological importance, since they carry out the most essential function in the economy of a living system — that is, the transmission and perpetuation of genetic information which controls the synthesis of proteins, and subsequently the synthesis of all other cellular components. Both DNA and RNA are composed of nucleotides which are mixtures of one mol of phosphate linked with one mol of sugar and one mol of a mixture of heterocyclic bases.

Those bases found in the nucleic acids belong either to the group pyrimidines the basal structure of which is a hexagonal ring with an aromatic character, containing two atoms of nitrogen (Figure 3.8), or to the group called purines, which consist of a pyrimidine ring fused to an imidiazole ring (*Figure 3.8*). Of the pyrimidines in DNA, the commonest found are thymine and cytosine;

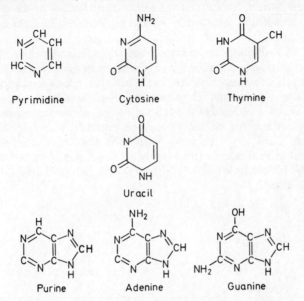

Pyrimidine   Cytosine   Thymine

Uracil

Purine   Adenine   Guanine

Figure 3.8. Illustrating structures of common pyrimidines and purines

the most abundant purines occurring in the nucleic acids are adenosine and guanine.

Two sugars occur in the nucleic acids, namely D-ribose in RNA, and D-2-deoxyribose in DNA. Both are pentagonal rings in which C3 and C5 are linked to a phosphate group, and C1 to the base in the mononucleotide structure. In the nucleic acids both pentoses exist in the furanose form, and the configuration about C1 is β:

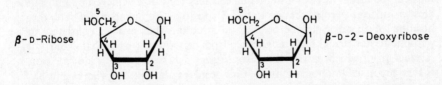

β-D-Ribose

β-D-2-Deoxyribose

Hence each monomer of nucleic acids consists of a nucleotide which is the outcome of the combination of one molecule of phosphate with one of a particular pentose and base. The combination of a pentose with a base constitutes

71

a nucleoside. Thus a nucleoside is a compound of a pyrimidine or purine base, and either D-ribose or D-2-deoxyribose – hence as indicated above the glycosidic linkage is the β-form, from N1 of the pyrimidines and N9 of the purines. Thus adenosine is adenine nucleoside which is 9-β-D-ribofuranosyladenine and deoxyadenosine is 9-β-D-ribofuranosyladenosine; guanosine is guanine nucleoside which is 9-β-D-ribofuranosylguanine, and deoxyguanosine is 9-β-D-deoxy-ribofuranosylguanine; thymidine is thymine nucleoside where thymidine is 1-β-D-2-deoxyribofuranosylthymine; cytodine is the cytosine nucleoside 1-β-D-2-deoxyribofuranosylcytosine. Thus from these nucleosides, nucleotides can be formed in which a phosphate is esterified to one of the pentose hydroxyls. In the ribonucleotides, the phosphate may be esterified with a hydroxyl in the carbon atom 2', 3' or 5' position, whereas in the deoxyribonucleotides one might expect to find esterification only with the 3' and 5' carbons, since these are the only free available hydroxyl sites. Two examples of nucleoside structures are shown in *Figure 3.9.*

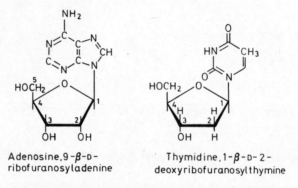

Adenosine, 9-β-D-ribofuranosyladenine

Thymidine, 1-β-D-2-deoxyribofuranosylthymine

Figure 3.9. Depicts two examples of nucleoside structures

Those nucleotide structures more commonly encountered as building units of DNA are illustrated in *Figure 3.10.* The formation of a polynucleotide requires the fusion of two or more nucleosides, and this is achieved by phosphate diester linkages formed between the 3' position of one pentose and the 5' position of an adjacent pentose unit; a structure thus evolves consisting of a phosphodiester backbone fabricated from alternating phosphate and sugar residues from which radiate the bases. *Figure 3.11* will serve to indicate how it is that successive monomeric nucleotide residues are linked, giving rise to 3', 5'-phosphodiester bridging between adjacent pentose nucleoside units.

Fundamentally then, DNA is composed of adenine, cytosine, guanine and thymine, and RNA of adenine, cytosine, guanine and uracil. The proportion and

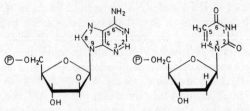

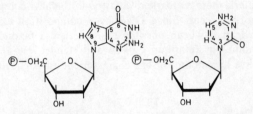

Figure 3.10. Depicts the main nucleotide structures figuring in the DNA molecule

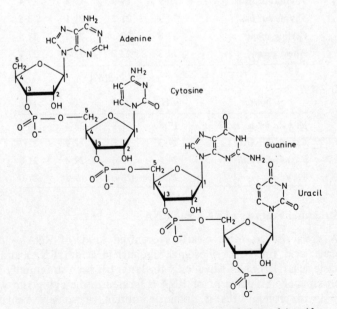

Figure 3.11. The polynucleotide structure of ribonucleic acid

composition of bases in DNA and RNA are seen to be characteristic of the host organism. For it has been shown that the base composition will vary according to the species of organism — in some adenine and thymine predominate in the DNA structure, and in other cytosine and guanine. But, thymine (T) will always be seen to pair exclusively with adenine (A), and guanine (G) with cytosine (C) in the DNA molecule. Hence $A \equiv T$; $G \equiv C$; $A + G \equiv C + T$ (Chargaff, 1950). In addition to this, the molar ratio of A/T and C/G is one. The composition of DNA has also been found to be essentially the same in all vertebrates — that is 40 mol per cent G + C, and 60 mol per cent A + T; whereas in the bacteria there is a marked variation in the base composition — from about 25–70 mol per cent. Since DNA base composition exhibits a widely stable characteristic it has been possible to formulate a taxonomy of bacteria based on DNA homology (Marmur, Falkow and Mandel, 1963). Table 3.4 lists a few representative base compositions characteristic of particular organisms or species.

TABLE 3.4
The Variation in Base Composition of DNA and RNA

Source	DNA			
	(A)%	(G)%	(C)%	(T)%
Thymus of man	30·9	19·9	19·8	29·4
Thymus of calf	29·9	21·2	21·1	28·5
Herring sperm	27·8	22·5	20·7	27·5
Wheat germ	27·3	22·7	22·8	27·1

Source	RNA			
	(A)	(G)	(C)	(U)
Mitochondria (rat liver)	17·8	31·8	28·4	20·9
Aspergillus niger	24·3	30·1	22·9	22·7
Escherichia coli	24·6	31·6	22·8	21·0
TMV	29·0	26·0	19·0	27·0

Some Structural Features of DNA and RNA

DNA gives rise to three clearly recognized kinds of RNA — messenger ribonucleic acid (mRNA), ribosomal ribonucleic acid (rRNA) and transfer ribonucleic acid (tRNA), of which only the latter has been structurally defined.

Approximately 80 per cent of RNA is located in the cytoplasm where it is associated with protein — that is, ribonucleoprotein (ribosomes). A small portion of RNA is also to be found in the nucleus (mRNA).

These three forms of RNA have been found to differ in base ratio, molecular size and function. Primarily, however, all three are intimately involved in the mechanics of protein synthesis which will be discussed later. Although their precise structures have not yet been determined — with the exception of that of tRNA — nevertheless it is reasonable to assume from existing experimental data that RNA is a polymer consisting of a ribose-3'-phosphate, 5'-ribose-3'-phosphate backbone, with the bases attached to the $C'1$ pentose unit (*Figure 3.11*).

In general all viral and animal DNA and RNA occur in close association with a basic protein — either a protamine or a histone. In somatic cells it is believed the nucleohistones are involved in the control of transfer of information from the DNA molecule to the protein-synthesizing system.

It is believed that mRNA is formed on the template of DNA and thereby transports the informational transcript to the ribosome surface — the protein assembly site. This information, inherent within the DNA complex in the form of a base pair sequence, is extracted by the mRNA in the negative form — thus at the ribosome site it functions as a template awaiting the arrival of a complementary amino-acid residue, the anticodon (*see* Chapter 6). A polyribonucleotide, rRNA is found predominantly in the form of ribosomes (40 per cent RNA and 60 per cent protein); tRNA is also a polyribonucleotide, and only recently has its structure come to light — namely that of ala-tRNA. One of the three suggested schemes of structure for ala-tRNA is shown in *Figure 3.12*. It will be seen that the molecule contains a relatively large proportion of unused bases as characterized by Me and H_2U, which are not to be found in other nucleic acids, H_2U refers to the fact that uracil possesses two or more atoms of hydrogen, Me refers to an additional methyl grouping and embraces such bases as 5-methylcytosine and 6-methylaminopurine. Furthermore, tRNA is characterized by possessing identical terminal sequences CCA (cytidyl-cytidyl-adenylic acid). The function of tRNA is to locate specific amino acids attached to its CCA terminals, on the mRNA template at the ribosome site. (*See* Protein Synthesis in Chapter 6, for a more detailed discussion of this topic).

Deoxyribonucleic acid consists of two right-handed helices entwined about a common axis, and held together by hydrogen bonds between the purine and pyrimidine bases which are at right angles to the axis of the helices (Wilkins, 1953; Watson and Crick, 1953). The phosphate-sugar complexes form the external part of the helix, with specific pairing of the bases, occurring as already stated at right angles to the long axis — thus forming the internal structure. A segment from the DNA molecule is shown in *Figure 3.13*, indicating two complementary pairs of bases with the hydrogen bonds in between. The nucleotides are linked by 3':5'-phosphate diester linkages. Also, pairing is highly specific — A with T and G with C. The nucleoside pairs at the helix interior measure 11 Å from C' to C'; furthermore the total diameter of the resultant molecule is somewhere in the region of 20 Å, with a distance of 3·4 Å between vertical base pairs.

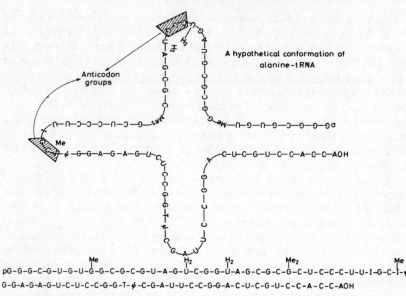

Figure 3.12. Suggested structure of ala-tRNA

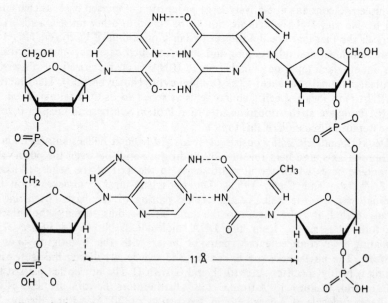

Figure 3.13. Depicts a segment of a DNA molecule showing two complementary pairs of bases, indicating the hydrogen bonds existing between them

THE CARBOHYDRATES

The carbohydrates are composed of carbon and water with a general formula of $C_n (H_2O)n$, and can be divided into four groups — the monosaccharides, disaccharides, polysaccharides and mucopolysaccharides. The prime biological importance of the carbohydrates is that they represent a source of energy in both plants and animals. Also in the plants they serve a function of structural support.

Fundamentally the stereoisomerismic carbohydrate structure is dependent on the quadrivalent property of carbon — since the four valencies are given specific direction in space. Carbon has a mass number of six, hence it follows that in the ground state the electron configuration will be Is^2, $2s^2$, $2p^2$ (thus meeting the requirements of the Aufbau principle). But carbon can be easily raised to an excited state, in which one of the 2s electrons is promoted to a 2p orbital — by a process known as sp^3 hybridization — whereby the distance of the orbitals from each other will be at a maximum. Thus they are directed towards the angles of a regular tetrahedron, such that the four bonds are separated by angles of $109^\circ \, 28'$. The four equivalent sp^3 orbitals are thus available for bond formation.

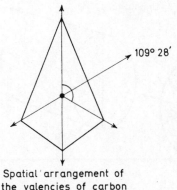

Spatial arrangement of
the valencies of carbon

Optical Isomerism

A light ray is a form of electromagnetic radiation composed of a number of energy waves pulsating at all angles to the direction of the ray's path. Should this ray pass through a section of an Iceland spar crystal, it can be demonstrated that the energy waves are all orientated in one particular direction. This type of light is called plane polarized light. The angle of the polarized light may be determined by allowing the light path to pass through another crystal lattice,

77

since if this complements the angle of the polarized beam of light, the light will be transmitted. Hence if an optically active solution is interposed between the two crystals the angle of the second crystal will have to be altered to permit transmission. Hence the optical activity of a compound of particular concentration can be determined using a polarimeter which will measure the change in degrees of specific rotation (α) using the formula:

$$[\alpha]_D = \alpha/I.c$$

where (α) is the number of degrees — which equals the amount of tilting of polarized light — by the sample; I is the length of the sample chamber; c is the concentration of the sample and D the wavelength of polarized light employed. For a diagrammatic explanation of the use of a polarimeter *see Figure 3.14.*

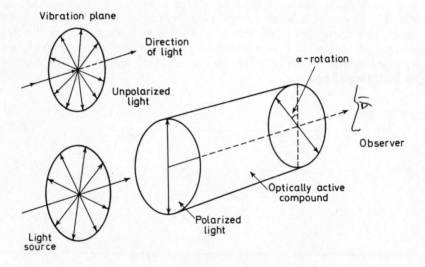

Figure 3.14. Schematic diagram of a polarimeter

If a compound rotates a plane of polarized light to the right, then the compound is said to be dextrorotatory (D+), and if rotation is in the opposite direction then it is laevorotatory (L−); in both cases the compounds are said to be optically active. If a molecule possesses the ability to rotate the plane of polarized light to the right its mirror image will rotate it equally to the left and these two optically active forms are called enantiomers. Such compounds generally possess identical physical and chemical properties. A 50–50 mixture of enantiomers is referred to as a racemic mixture.

The Monosaccharides

The monosaccharides contain 5 or 6 carbon atoms. It is known that in solution most carbohydrates exist in rings, and not as linear chains — though D+-glucose in solution behaves as though it were a linear form (*see* page 80). The basic ring structure is composed of one oxygen and five carbon atoms. But again, usually glucose and many other sugars are structurally a six membered ring of the pyranose type. The following depicts α-pyranose forms of some naturally occurring D-sugars:

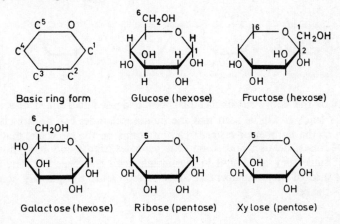

Basic ring form   Glucose (hexose)   Fructose (hexose)

Galactose (hexose)   Ribose (pentose)   Xylose (pentose)

Some compounds of fructose, and the pentoses exist as 5 membered ring structures — known as fructopyranose rings:

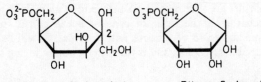

β-D-Fructose-6-phosphate   α-D-Ribose-5-phosphate

It will be seen from this that the carbohydrate structure is complicated further by the fact that D+-glucose can exist in two other forms known as α and β, which possess different optical activity. Both forms are interconvertible in solution, this being achieved by a process known as mutarotation.—

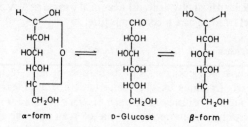

α-form   D-Glucose   β-form

The α and β forms of D-glucopyranose and fructopyranose are:

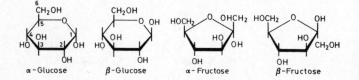

α-Glucose   β-Glucose   α-Fructose   β-Fructose

The units of the above substances may be condensed to form a disaccharide molecule. Hence it will be seen that the monosaccharides can be sub-classified according to the number of resident carbon atoms on the molecule; that is into trioses, pentoses, hexoses and so on. The pentoses, ribose and deoxyribose are important since they are found in the nucleic acids (*see* page 71). The hexose glucose is also of major importance since it represents the primary source of energy for the cell.

Disaccharides

The disaccharides are the condensation products of two monomers of monosaccharides. Their empirical formula is $C_{12}H_{22}O_{11}$. In plants the most important disaccharides are sucrose and maltose, and in animals, lactose.

Sucrose or saccharose is the sugar of cane or beets, and is formed by the combination of one molecule of glucose with fructose. It will be noticed that the molecule is a double glycoside, and is therefore a non-reducing sugar:

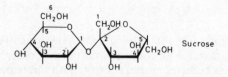

Sucrose

Hence formally, sucrose is known as -D-glucosyl- -D-fructoside.

80

The disaccharide maltose on the other hand is formed by the condensation of two α-glucose molecules resulting in an amalgam of the glucopyranose rings defined as α-D-glucosyl-1, 4-D-glucose:

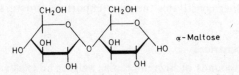

Maltose unlike sucrose is a reducing sugar, by virtue of its free aldehyde groups. Lactose, the sugar of milk (5 per cent content) is formed by the condensation of glucose and galactose:

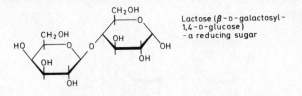

Polysaccharides

The polysaccharides result from the condensation of many monosaccharide units. Their empirical formula is $(C_6H_{10}O_5)_n$. Hydrolysis of the polysaccharides yields molecules of simple sugars. The most important polysaccharides are starch and glycogen which represent the reserve food substances of cells in plants and animals respectively.

Starch is a mixture of two polymers, amylose and amylopectins, both of which are composed of D-glucose units. Amylose consists of a linear unbranched structure composed of some 200–2 000 glucose units welded together by α-1, 4 glucosidic bonds. The amylopectin molecule on the other hand is a branched structure consisting of 200–5 000 units. The units of amylopectin are joined by α-1, 4 linkages and 1, 6 linkages occurring at the branch points:

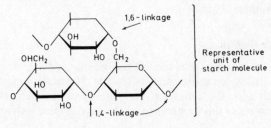

Glycogen is found only in animals. It is a highly branched polymer of glucose units, and is widely disseminated in animal tissues, particularly in liver cells and muscle fibres.

Cellulose is a linear polysaccharide chain composed of β-1, 4 linked glucose units. This polymer constitutes the most important structural element in plant cells.

The Mucopolysaccharides

These are compounds of high molecular weight and exist as a polysaccharide-protein complex in animal tissue where fundamentally they appear to play a structural role. Representatives of this group are chondroitin sulphate, mucoitin sulphate, heparin, hyaluronic acid and blood group substances.

Chondroitin sulphate. — This is found in cartilage and bone and consists of repeating units of:

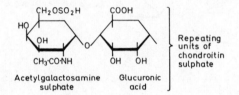

Acetylgalactosamine sulphate · Glucuronic acid · Repeating units of chondroitin sulphate

Mucoitin sulphate. — This mucopolysaccharide is found in saliva and consists of acetylglucosamine sulphate glucuronic acid units.

Hyaluronic acid. — This substance forms the cement of interstitial tissue, and is composed of alternate units of glucuronic acid and N-acetylglucosamine, linked together to form a thread-like structure.

Blood group substances. — These consist of 85 per cent carbohydrate and 15 per cent peptide, the two moieties being linked together covalently. The following sugars account for the carbohydrate content: 6-deoxy-6-galactose, D-galactose; N-acetyl-D-glucosamine; N-acetyl-D-galactosamine; and N-acetyl-neuraminic acid. With regard to the peptide moiety 15 L-amino acids have been identified (Watkins, 1966).

Carbohydrate Digestion

The object of digestion is to convert polyglucose and the other polysaccharides into smaller utility molecules such as glucose. Mammalian digestion commences in the mouth with the action on the starches of the enzyme, α-amylase. The enzyme commences this breakdown of the large glucose polymers by converting them to dextrans; this is achieved by the degradation of the 1,4 linkages. This enzyme α-amylase is present in the saliva which is a watery fluid containing the normal somatic ions in relatively small concentrations, in addition to bicarbonate buffer which maintains the pH at about 6–7. Also present in saliva are some heavily hydrated mucoproteins which aid lubrication of the masticated food. Salivary digestion terminates in the acidic environment of the gastric pouch.

The major part of carbohydrate digestion occurs in the small intestine and the various aspects of this are summarized in *Figure 3.15*. The efficiency of these enzymes in bringing carbohydrate digestion to its ultimate completion can be affected by a variety of conditions. Of these probably the most common – particularly in Negroes – is lactase deficiency, which can cause infantile

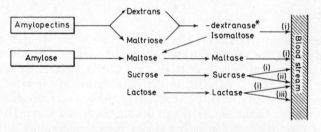

*α-dextranase degradates further the side chains of amylopectin
(i) glucose (ii) fructose (iii) galactose

Figure 3.15. Aspects of carbohydrate digestion

diarrhoea; this, if left unchecked can lead to malnutrition since these children will not be able to make use of milk nutritionally.

Following digestion the sugars are then absorbed into the blood stream. Some sugars are more readily absorbed than others. It appears that galactose is the most readily absorbed, followed by glucose, mannose, fructose and the pentoses in decreasing order of absorption. It has been shown that the specificity of active transport mechanism depends on the hydroxyl in the $C'2$ position. The active transport processes follow normal reaction kinetics. Hence the monosaccharides are the end-product of carbohydrate digestion of which glucose is the principal molecule. Following absorption the monosaccharides are utilized in the following way: (1) immediate utilization, (2) glycogen storage; and (3) conversion to fat (*see* Chapter 8).

LIPIDS AND STEROIDS

The lipids are esters of the higher aliphatic alcohols, all of which are insoluble in water, but soluble in solvents such as benzene, chloroform and ether. This property of insolubility is directly attributable to the presence of long aliphatic hydrocarbon chains which are hydrophobic and non-polar by nature.

The lipids can be classified as follows:

```
                              Lipids
                                 |
        ┌────────────────────────┼────────────────────────┐
     Lipoids              Conjugated (complex)           Simple
    ┌────┴─────┐          ┌────────┴────────┐        ┌─────┼─────┐
 Sterols   Steroids  Phospholipids  Sphingolipids  Fats  Oils  Waxes
```

Simple Lipids

The simple lipids are alcohol esters of fatty acids; the series is composed of the fatty acids, the neutral fats, the oils and the waxes. The saturated fatty acids have an empirical formula of $C_nH_{2n+1}.COOH$; whereas the unsaturated fatty acids are $C_nH_{2n-1}.COOH$. In the classification of the fatty acids the suffix anoic refers to a saturated compound; an enole is a compound containing one double bond; dienoic indicates two double bonds and so on. To mark the position of these resident double bonds, the present convention is to employ the symbol Δ. In the case of Δ^{9-10}-octadecanoic (oleic) acid, the double bond occurs between the ninth and tenth carbon atom. Oleic acid ($C_{17}H_{33}COOH$) has the structure:

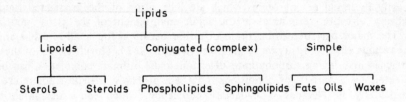

Some of the common fatty acids are listed in Table 3.5.

TABLE 3.5
Common Representatives of the Fatty Acid Series

Name	Formula	Systematic name	Natural source
Caproic	$C_5H_{11}.COOH$	Hexanoic	Palm and coconut oils
Capric	$C_9H_{19}.COOH$	Decanoic	..
Palmitic	$C_{15}H_{31}.COOH$	Hexadecanoic	Animal and plant oils
Stearic	$C_{17}H_{35}.COOH$	Octadecanoic	..
Oleic	$C_{17}H_{33}.COOH$	Δ^{9-10}-octadecanoic	..
Arachidic	$C_{19}H_{39}.COOH$	Eicosanoic	Peanut oil

LIPIDS AND STEROIDS

The neutral fats and oils are triesters of fatty acids and glycerols. The composition of fats and oils derived from animals will vary according to the constituents of the host's diet. For instance butter is a mixture of glycerols of butyric, oleic, palmitic, stearic, arachidic and others and by way of comparison plant fats and oils are relatively constant in composition. Oils have a greater proportion of unsaturated acids and are consequently liquid at 20°C whereas the fats are solid at this temperature. Waxes on the other hand have a higher melting point than the neutral fats. The waxes are esters of higher fatty acids and alcohols – consequently they are immiscible with water and very difficult to hydrolyse.

To summarize briefly – all the fatty acids mentioned are seen to possess an even number of carbon atoms in their backbone structure. Also most of the naturally occurring fatty acids have a higher molecular weight. Those with four carbons or less are soluble in water whereas those with 10 or more carbon atoms in addition to being insoluble are also solid at 20°C.

Complex Lipids

This group includes the phospholipids and the glycolipids and both are seen to be primarily concerned in the structural activities of the cell system – particularly in relation to the architecture and properties of cell membranes (*see* Chapter 4). The phospholipids are diesters of phosphatidic acid and contain several important lipid complexes, these being phosphatidyl lecithins, ethanolamines and the serines. The major role of the phospholipids is the maintenance of both the integrity of the oxidative phosphorylation processes occurring in the mitochondrion, and the status of the cytoplasmic membrane, thereby influencing ion transportation and regulation of membrane permeability. The phospholipids are often called polar lipids, since they have attached to one end of the long hydrocarbon chain a polar group which endows the structure with hydrophilic properties, thus rendering it capable of binding water by hydrogen bond formation.

The glycolipids are probably best characterized by the sphingolipids which are a collection of lipids, in which glycerol has been replaced by the long-chain, amino alcohol, sphingosine. There are three principal groups of sphingolipids: sphingomyelins, cerebrosides and gangliosides. The sphingomyelins, on hydrolysis, yield sphingosine, choline, phosphate and fatty acid.

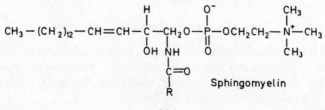

Sphingomyelin

85

Sphingomyelin is to be found in nerve tissue, where it forms part of the sheath of a myelinated axon.

The cerebrosides are characterized by the fact that they will yield on complete hydrolysis, sphingosine, fatty acids and a carbohydrate — usually galactose.

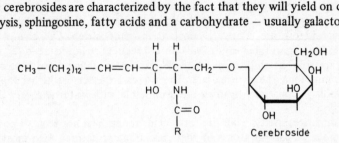

In certain pathological conditions the above sugar moiety, galactose, is replaced by glucose. There are a number of conditions associated with an abnormal deposition of lipid. For instance in Niemann–Pick disease there is an increase in deposition of sphingomyelins and gangliosides in both the brain and liver. Similarly in Gaucher's disease there is an increased deposition in the liver of the cerebroside kerasin. Furthermore gargoylism is due to the deposition of an abnormal mucopolysaccharide in the tissues (Thompson and King, 1964).

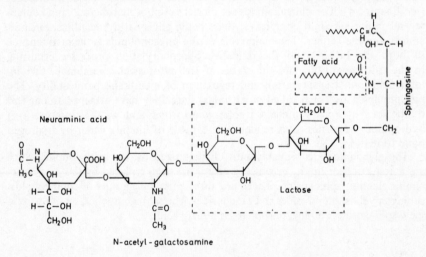

Figure 3.16. Molecular structure of a ganglioside

Finally the gangliosides, which are complex sphingolipids, contain sphingosine, fatty acids, carbohydrates and neuraminic acid. Together the sphingolipids account for the majority of the lipid content of the brain. The molecular structure of a ganglioside is outlined in *Figure 3.16*.

The Steroids

The steroids are lipids derived from the cyclopentanoperhydrophenanthrene nucleus:

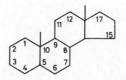

Fundamentally their biological activity is dependent on the presence of attached groups, and their spatial relationship to the above nucleus. The steroids include a series of important structures capable of being synthesized by the mammalian system, and other organisms, which play an integral role in a variety of physiological and biochemical activities (*see* Fieser and Fieser, 1959).

The predominant steroid in man is cholesterol. The cholesterol molecule is in fact a sterol: that is, it possesses a double bond between $C'5$ and $C'6$. Cholesterol can also be classed as a polar lipid, although its polar group (-OH) is only weakly hydrophilic.

Cholesterol synthesis can occur in almost any tissue. Acetate is in fact the smallest molecule which can enter into cholesterol synthesis. The cholesterol synthesis pathway commencing with the activation of acetyl coenzyme A is illustrated in *Figure 3.17*. As indicated in this illustration, cholesterol can be synthesized in a variety of different tissues – for example, in skin. It is believed that squalene is the control point for the skin synthesis of cholesterol.

Examples of those steroids vital to the human machine include the bile acids, the adrenocortical hormones and the sex hormones.

Bile contains three important bile acids, lithocholic, deoxycholic and cholic acid which are excreted as conjugates with glycine and taurine yielding the emulsifying agents glycholic and taurocholic acid.

Whereas the synthesis of cholesterol is widely disseminated in the body, the conversion of this precursor substance to the steroid hormones in appreciable quantities is limited to the adrenal cortex and the endocrine portions of the sex organs (*Figure 3.18*). The adrenocortical hormones elaborated by the adrenal cortex are chiefly concerned with the regulation of protein and carbohydrate metabolism and electrolyte balance in which aldosterone, corticosterone, deoxycorticosterone and cortisol are the prime participants. Cortisol is the main glucocorticoid in man, sheep and monkey. The processes which are influenced by this hormone are glucogenesis, the life span of lymphocytes, and the inflammatory processes. Aldosterone is the mineralcorticoid of major importance in man and most other mammals. It controls sodium retention (*see* Homoeostasis).

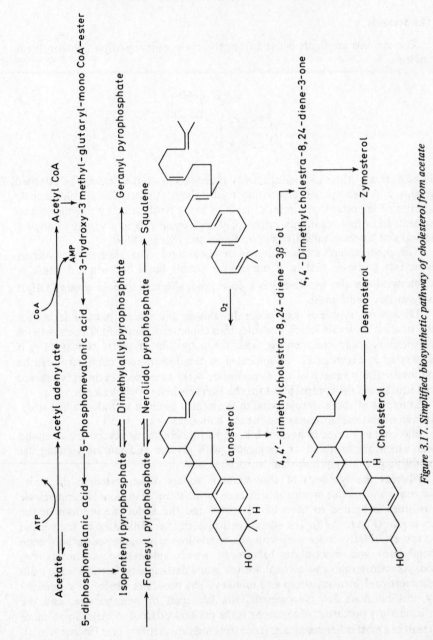

Figure 3.17. Simplified biosynthetic pathway of cholesterol from acetate

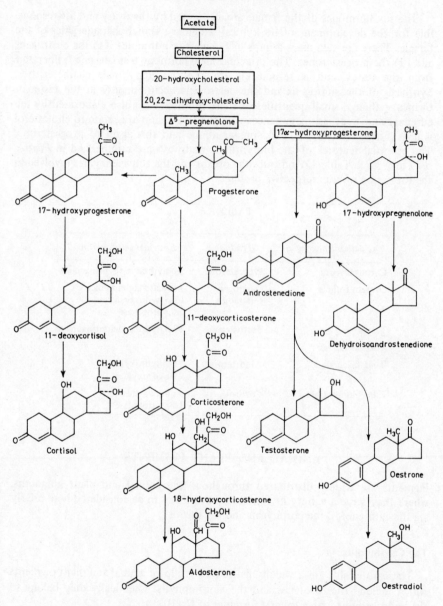

Figure 3.18. Biosynthesis of the steroid hormones (simplified)

The sex hormones of the female are elaborated by the ovary and are responsible for the development of the typical secondary sexual characteristics of the female. There are two main groups of female sex hormones: (1) the oestrogens and (2) the progesterones. The principal male hormone testosterone is liberated from the testes, and is responsible for the secondary male characteristics. Synthesis of the androgens and the oestrogens occurs mainly in the testes or ovaries — though small quantities of these hormones are also elaborated by the adrenal gland. Synthesis commences with the conversion of acetate to cholesterol which is then converted to Δ^5-pregnenolone and this again to progesterone. The essential features of the biosynthetic pathways are summarized in *Figures 3.17* and *3.18*. Table 3.6 indicates the organs and the active enzymes involved in the formation of each respective product.

TABLE 3.6

The elaborating organ	Product	Enzymes participating
Corpus luteum	Progesterone	3-β-hydroxydehydrogenases
Ovarian follicle	Oestrogens Oestradiol	Aromatizing enzyme 17-β-hydroxysteroid-dehydrogenase
Testes	Testosterone	17-β-hydroxydehydrogenase
Adrenal Cortex		
Zona fasciculata	Aldosterone	11-β-hydroxylase 21-hydroxylase
Zona glomerulosa	Cortisol	As above plus 18-hydroxylase

ANIMAL AND PLANT PIGMENTS

Pigments are widely distributed throughout the animal and plant kingdoms, where they serve a variety of functions, and those to be considered here briefly are the carotenoids, the porphyrins and the flavins.

The Carotenoids

The carotenoids are a widely distributed group of animal and plant pigments which include the carotenes and the xanthophylls. Chemically they belong to the hydrocarbons with a general formular of $C_{40}H_{56}$.

The carotenes exist in three forms the α, β and γ carotenes. Apart from functioning in the photosynthetic mechanism of plants the carotenes serve as a source from which animal tissue can obtain Vitamin A. For the hydrolysis of β-carotene will yield 2 molecules of vitamin A and 1 molecule following the hydrolysis of the α form. The carotenes impart to the tissues of vegetables their characteristic yellow colouring — whereas the carotenoid lycopene is the responsible colouring agent of ripe tomatoes.

Related to the carotenes are the xanthophylls, of which lutein is the most common example. Lutein is to be found in close association with the chloroplasts of green leaves, and is responsible for the autumnal colouration of leaves following the diminishing presence of chlorophyll at that time.

The Porphyrins

Like the carotenes, the porphyrins are widely disseminated in nature, and structurally they are variations on the tetrapyrrole theme. The most important representatives of this type of pigment are the haemochromes and the chlorophylls.

The haemochromes are chromoproteins in which the prosthetic group occurring in the vertebrate organism is iron-containing haem — whereas in the arthropods, for example, the prosthetic group is a copper-containing thiopeptide. In principle these pigments perform the respiratory function which is required to form oxyderivatives whereby oxygen can be transported within the organism (*see* Quaternary Structure of Protein, page 68).

By contrast, chlorophyll which is the principal photosynthetic pigment is a non-protein molecule. Fundamentally the chlorophyll molecule consists of a tetrapyrrole frame with a magnesium nucleus. Those organisms performing oxygen-evolving photosynthesis are seen to contain at least one kind of chlorophyll — that is chlorophyll a ($C_{55}H_{72}O_5N_4Mg$) which is sometimes, and particularly in the higher green plants, supplemented by a second type of chlorophyll, designated chlorophyll b ($C_{55}H_{72}O_6N_4Mg$). Other kinds of supple-

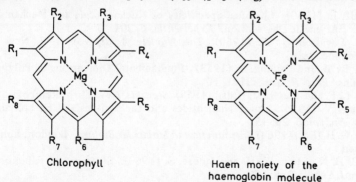

Chlorophyll

Haem moiety of the haemoglobin molecule

Figure 3.19. Diagram depicting comparative structures of chlorophyll and haem

91

mentary chlorophylls c, d and e also occur, and are to be found in the lower organisms such as the photosynthetic bacteria, algae and diatoms. Those pigments associated with photosynthesis – the chlorophylls, carotenoids and phycobilins – will be discussed further in Chapter 5, page 153. In *Figure 3.19* a diagram is shown comparing the basic structure of chlorophyll with that of haemoglobin.

The Flavins

The flavins are yellow nitrogenous pigments often bound as a prosthetic group to a protein. As flavoproteins they function as coenzymes in important dehydrogenation reactions. Essentially all the flavins may be represented by the molecule riboflavine (vitamin B_2):

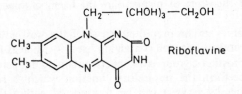

There are flavin coenzymes – flavin mononucleotide (FMN) and flavin adenine dinucleotide (FAD).

BIBLIOGRAPHY AND REFERENCES

Chargaff, E. (1950). 'Chemical Specificity of Nucleic Acids and Mechanisms of their Enzymatic Degradation.' *Experientia*, **6**, 201.
– and Davidson, J. N. (1960). *The Nucleic Acids*, vols 1–3. New York; Academic Press.
Corey, R. B. and Pauling, L. (1953). 'Fundamental Dimensions of Polypeptide Chains.' *Proc. R. Soc.*, B**141**, 10
Fieser, L. F. and Fieser, M. (Eds.) (1959). *Steroids*. New York; Rembold.
Ginsburg, V. (1964). 'Sugar Nucleotides and the Synthesis of Carbohydrates.' *Adv. Enzymol.*, **26**, 35.
Haggis, G. H. (Ed.) (1964). *Introduction to Molecular Biology*. London; Longmans Green.
Hirs, C. H. N., Stein, W. H. and Moore, S. (1960). 'Structure of Ribonuclease.' *J. biol. Chem.*, **235**, 633.
Kauzmann, W. (1959). 'Some Factors in the Interpretation of Protein Denaturation.' *Adv. Protein Chem.*, **14**, 1.

BIBLIOGRAPHY AND REFERENCES

Kendrew, J. C., Bobo, G., Dintzis, H. M., Parrish, R. C., Wyckoff, H. and Phillips, D. C. (1958). 'A Three-dimensional Model of the Myoglobin Molecule obtained by X-Ray Analysis.' *Nature, Lond.*, **181**, 662.

Klotz, I. (1957). 'Some Principles of Energetics.' In *Biochemical Reactions*. New York; Academic Press.

– (1960). 'Non-covalent Bonds in Proteins Structure.' *Brookhaven Symp. Biol.*, **13**, 25.

Marmur, J., Falkow, S. and Mandel, M. (1963). 'New Approaches to Bacterial Taxonomy.' *A. Rev. Microbiol.*, **17**, 329.

Morowitz, H. J. (1958). *Microsomal Particles and Protein Synthesis*. Ed. by R. B. Roberts. Oxford; Pergamon Press.

Pauling, L. (1960). *The Nature of the Chemical Bond, 3rd ed.* Ithaca, New York; Cornell University Press.

– Corey, R. B. and Branson, H. R. (1951). 'The Structure of Proteins: Two Hydrogen Bonded Helical Configurations of the Polypeptide Chain.' *Proc. natn. Acad. Sci., U.S.A.*, **37**, 205

Perutz, M. F. (1962). *Proteins and Nucleic Acids*. Amsterdam; Elsevier.

– (1964). 'The Haemoglobin Molecule.' *Scient. Am.*, **211**, 64.

– and Lehmann, H. (1968). 'Molecular Pathology of Human Haemoglobin.' *Nature, Lond.*, **219**, 902.

– Muirhead, H., Cox, J. M. and Goaman, L. C. G. (1968). 'Three Dimensional Fourier Synthesis of Horse Oxyhaemoglobin at 2·8 Å Resolution: Atomic Model.' *Nature, Lond.*, **219**, 131.

– Rossmann, M. G., Cullis, A. F., Muirhead, H., Will, G. and North, A. C. T. (1960). 'Structure of Haemoglobin: A Three Dimensional Fourier Synthesis at 5·5 Å Resolution Obtained by X-Ray Analysis.' *Nature, Lond.*, **185**, 416.

Pontecorvo, G. (1952). 'Genetic Formulation of Gene Structure and Gene Action.' *Adv. Enzymol.*, **13**, 121.

Sanger, F. (1952). 'The Arrangement of Amino Acids in Proteins.' *Adv. Protein Chem.*, **7**, 1.

– (1956). 'The Structure of Insulin.' In *Currents in Biochemical Research*. Ed. by D. E. Green. New York; Interscience Publishers.

Smyth, D. G., Stein, W. H. and Moore, Y. (1963). 'The Sequence of Amino Acid Residues in Bovine Pancreatic Ribonuclease: Revisions and Confirmations.' *J. biol. Chem.*, **238**, (1), 227.

Thompson, R. H. S. and King, E. J. (1964). *Biochemical Disorders in Human Disease*. London; Churchill.

Watkins, W. M. (1966). 'Blood Group Substances.' *Science, N. Y.*, **152**, 172.

Watson, J. D. and Crick, F. H. C. (1953). 'Molecular Structure of Nucleic Acids: A Structure for Deoxypentose Nucleic Acids.' *Nature, Lond.*, **171**, 737.

Wilkins, M. H. F. (1953). 'Molecular Structure of Deoxypentose Nucleic Acids.' *Nature, Lond.*, **171**, 738.

4 – Organization of Molecular Components

(The cell wall – The unit membrane – Cytoskeletal membranes: structure and function

THE CELL WALL

Prior to reviewing the structure of the cell membrane, the extraneous coat will be considered. In many animals this structure is hardly discernible, but in the cells of plants and in particular in groups of the protista (where it is a very prominent feature of the general morphology), the delicate cytoplasmic membrane is surrounded by a thick structure with a different chemical composition, and this is known as the cell wall.

Fundamentally the function of this extraneous coat is to maintain the microenvironment of the cell – for, so far as is known, the cell wall is not endowed with semipermeable properties; in consequence it does not participate in the selective transportation of metabolites in or out of the cell and it appears also to be devoid of enzymatic function. Hence the cell wall is a protective, supporting structure, the essential function of which is seen to be the prevention of swelling and osmotic lysis of the protoplast as a result of exposure to a hypotonic environment.

Principally the primary structure of a cell wall is a macromolecular lattice-work, possessing a high tensile strength and having a chemical composition which is generally of a polysaccharide nature. In plants and the green algae the structural polysaccharide is cellulose, a glucose polymer, and in the fungi it is chitin which is a polymer of acetylglucosamine; in the lower protista the structural material is a mucocomplex, a heteropolymer composed of two amino-acid units – acetylglucosamine and acetylmuramic acid.

In plants the cell wall consists of two major components, a microfibrillar portion which is embedded in a continuous amorphous matrix. The chemical

94

composition of the microfibrillar element is usually cellulose which is a polymer of D-glucose. The matrix on the other hand is composed largely of hemicelluloses and pectin substances. Generally in a plant cell wall the matrix is composed of one or more predominating substances of the hemicellulose group such as the xylans, or the galactans and arabogalactans. The pectin substances are assembled from the polymerization of α-1,4-D-galacturonic acid molecules. The micro-fibrils are orientated and interlinked with the matrix thereby imparting to the total complex, which is an orderly lattice, a distinctive property of rigidity. The plant cell wall consists of a thin primary wall, which is formed first, and a thicker secondary wall — the two being separated by a space called the middle lamella. In plant cells pectin is found in high concentration in the middle lamella, and the exterior primary wall, and cellulose is seen to be the predominant structural material throughout.

As already stated the mucocomplex substance is primarily that element conferring structural rigidity on the cell walls in the lower protista, such as the bacteria. In fact micro-organisms without the extraneous layer become fragile spheres bounded by the semipermeable cell membrane. Those cells grown in a medium which prevents cell wall synthesis are called protoplasts; these, if placed in distilled water, will lyse as a result of osmotic swelling. Those cells still in possession of their cell walls, however, do not experience this osmotic trauma. Essentially, then, the cell wall consists of a fibrous mucopeptide portion and a matrix composed of teichoic acids.

The chemistry of the mucopeptide portion of the cell wall is complex. Basically the mucopeptides are laid down along carbohydrate strands which consist of polymerized dimers of N-acetylglucosamine and N-acetylmuramic acid linked to one another by glucosidic bonds:

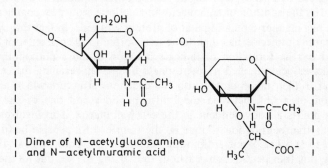

Dimer of N—acetylglucosamine
and N—acetylmuramic acid

For example, in the staphylococcus cell wall peptide growth occurs from the carboxyl group attached to the N-acetylmuramic acid unit; the peptide consists of alanine, D-isoglutamine, lysine and D-alanine which terminates with the assembly of five glycines which function as the stitching—knitting together the structure of the carbohydrate backbone. It is interesting to note that it is at these growing points that antibiotics such as pencillin act.

Biosynthesis of the cell wall commences with the UDP-sugars in the cell

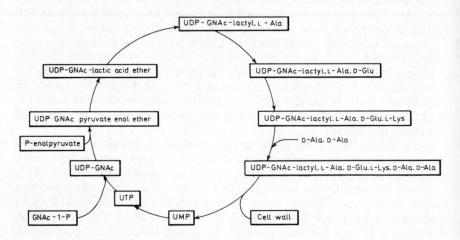

Figure 4.1. Indicating biosynthesis pathway of the UDP-muramic pentapeptide cell wall precursor. The individual amino acids are added utilizing ATP (after Strominger and Tipper, 1965)

cytoplasm (*Figure 4.1*). UDP-N-acetylgalactosamine then reacts with phospho-enolpyruvate to produce an unstable linkage at the $C'3$ position, which is then reduced by NADH, resulting in the formation of muramic acid. This is then followed by the addition of amino acids in a definite sequence, commencing with alanine, in a manner simulating that of protein synthesis. In fact this cytoplasmic sugar peptide molecule has then to be transported to the cell exterior, this being achieved by the formation of linkages between it and a phospholipid carrier, such linkages being termed pyrophosphate bonds. On reaching the cell exterior the sugar is then incorporated into the backbone structure which requires the formation of a link between the bridge peptides and one of the D-alanine growing points already resident in the cell wall matrix. This involves a process known as transpeptidation — that is, the transfer of a peptide bond from one location to another. Some antibiotics mediate at this point; that is, they act by blocking this transpeptidation reaction, thus fixing the peptide on the phospho-lipid carrier and preventing its subsequent transfer.

The matrix is mainly composed of teichoic acids, which in general are polymers of ribotol or glycerols held together by phosphodiester linkages. In fact much of the antigenicity of an organism is resident in this portion of the cell wall. The antigenic specificity, which is of great importance in the diagnostic and epidemiological microbiology of bacteria, is an expression of their specific surface structures. Those antigens designated O or somatic are located in the cell wall. The essential differential chemical features of the Gram-positive and Gram-negative bacteria cells are summarized in Table 4.1.

TABLE 4.1

Essential Features of Chemical Components of the Eubacterial Cell Wall

	Lipids	Proteins	Polysaccharides	Mucocomplex substances	Teichoic acids
Structural residues	Composition uncertain	Amino acids	Simple mono-saccharides or amino sugars	Glucosamine Nuramic and glutamic acid Alanine Diaminopimelic acid	Ribotol phosphate Glucose
Gram -ve eubacteria	+	+	+	+	−
Gram +ve eubacteria	−	−	+/−	+	+/−

Some Examples of the Chemica Composition of Bacterial Capsules

		Gram positive			Gram negative	
Class of capsular material		Leuconostoc	Pneumococcus	B. anthracis	E. coli	Acetobacter group
	Structural unit	Glucose	Amino sugars Uronic acids	D-glutamic acid Amino sugars	Galactose Uronic acids	Galactose Glucuronic acid
		Polysaccharide (Dextran)	Polysaccharide	Polypeptide and polysaccharide	Poly-saccharide	Poly-saccharide

The eubacteria also produce a variety of capsular substances which are predominantly polysaccharide by nature — from observation this indicates that the capsule is not an essential cell component, but rather a structure enhancing survival of the cell *in vivo;* as a consequence its presence is associated with virulence. Indeed, it has been shown that the capsule makes the organism less susceptible to phagocytosis (*See* Chapter 6, page 192 and Chapter 7, page 229).

In animal cells, polymers of glycoproteins and polysaccharides in the form of hyaluronic acid are found, together with other polysaccharides such as collagen which function principally in a structural capacity, as in connective tissue. As with bacteria, the cell surfaces of mammalian tissue have been specifically determined genetically, this serving to segregate a tissue species, and thus maintain its integrity. This is exemplified in the mammalian blood group substances, which are to be found bound to the surfaces of erythrocytes, and which contain different amino acids and carbohydrates, by virtue of which it is possible to identify and classify the human population into the four blood groups, A, B, O and AB.

A biochemical explanation of the differences between blood group A and B individuals is that the group A individual possesses the following antigenic determinant on the erythrocyte:

*N—acetylgalactosamine—1,3—galactose—β—1,3—N—acetylglucosamine
$|$
β—1,2 (usually)
$|$
Fucose

and the only difference between this and B substance is that in the latter there is no N-acetyl group on the galactosamine indicated by the above for a group A individual. Furthermore, A and B substances can be converted to H substance simply by removing the N-acetyl*-sugar. Also if fucose is removed from H substance then Lewis-a (Le^a) is produced (*see* page 278).

THE UNIT MEMBRANE

In Chapter 3 the main molecular components encountered in a living system were briefly reviewed. This chapter will describe the manner in which these components have become aggregated to form functional macromolecular assemblies.

In the mammalian organism membranes serve to subdivide the total cellular complex into functional compartments, thereby separating the extracellular fluids of the organism from the intracellular, and their aggregate fluids. The boundary membrane of a cell, often referred to as the plasma membrane or plasmalemma, is seen to possess definite physicochemical properties capable of maintaining a concentration gradient between the internal and external environment — thus regulating the translocation of molecular metabolites between the tissue compartments. Furthermore cells possess intracellular compartments which perform specialized functions the integrity of which are maintained by a particular intracellular membrane structure. For instance in the neurone, the division between the intracellular and extracellular fluids by a specialized membrane system forms the basis of the propagation of a nerve impulse — for, following the redistribution of membrane ions, the bio-electrical status of the membrane is transiently altered, resulting in an increase in local permeability of the axon membrane, and the initiation of an action potential. It will become evident that the integrity of a membrane — the selective permeability properties and ability to function as an intermediary boundary between the internal and external environment of which is primarily centred round the order in which its component lipoproteins are assembled (thereby endowing the total structure with a low permeability to ions and a high permeability to lipid soluble material) — is regulated in part by neural and hormonal activity (*see* Homoeostasis, page 288).

The plasma membrane is beyond the resolution of the ordinary light microscope, thus evidence concerning membrane structures, and the resultant comprehension of them, is based on chemical analyses, x-ray diffraction studies and electron microscope techniques together with data assembled from the study of such material as monolayer films and myelin figures. Contemporary knowledge with regard to the chemical composition of membranes, has evolved largely from information gathered from the studies of erythrocytic stroma, mammalian myelin

TABLE 4.2

The Chemical Composition of Membranes (as % of Total Dry Weight)

Type	Phospholipids						Cholesterol	Protein
	Phosphatidic acid	Phosphatidyl-ethanolamine	Phosphatidyl-choline	Phosphatidyl-serine	Phosphatidyl-inositol	Sphingolipids		
Human myelin sheath	—	∿14	∿10	∿5	0	∿30	25	40
Erythrocyte plasma membrane	<5	20	∿23	11	2	18	25	60
Mitochondrial membrane	0	∿28	∿48	0	∿8	0	5	60

and bacterial membranes (*see* Table 4.2). For instance following the haemolysis of a red cell in a hypotonic solution the resultant debris is the cell stroma which is composed almost exclusively of the plasmalemma, consisting of a lipid—protein ratio of 1:1·7. Robertson (1961), working with myelin membranes, proposed the concept of a unit membrane. On the basis of evidence from the electron microscope Robertson postulated the universal presence of a trilamina membrane structure. In the electron micrographs this three-layered arrangement appears as two parallel dark lines separated by a light space. He concluded that the two electron-dense lines represented the polar ends of phospholipid molecules together with associated protein molecules. The Robertson unit membrane proposes a basic morphology occurring universally at the surfaces of all plants, animals and protozoan cells. Fundamentally then, the topography of the boundary membrane may be considered to be a closed vesicular or tubular system with a structural rim consisting of a mosaic of repeating units of a lipoprotein nature; this phase is continuous with an equally complex and tortuous cytoplasmic one which forms the external phase of the cytoplasmic membrane and the endoplasmic reticulum.

The actual organization of the chemical components of the membrane into a formalized and dynamic structure has been the subject of intensive scientific research and polemics for the last 40 years, and is still not completely resolved. Essentially it is accepted without reservation that biological membranes are complex lipoprotein structures. Membrane proteins can be classified separately and exclusively and seen to possess the following properties: (1) the total lack of disulphide groups, which most probably accounts for the flexibility of such proteins, and that ability to alter their conformation which is the prime requisite for their function in active transport; (2) a hydrophobic composition; (3) an acidic nature, since they have a very high glutamic and aspartic acid content; and (4) amino-acid sequences leading to a tertiary structure suitable for lipoprotein membrane formation.

The number of lipids associated with a membrane structural protein is limited, so it seems, by the inherent nature of the hydrocarbon chains of the lipid. Five phospholipids have been identified, these being phosphatidylcholine, phosphatidylethanolamine, phosphatidylinositol, phosphatidylserine and phosphatidic acid. Recent work on the salt glands of aquatic birds has revealed that the phospholipids are not simply inactive structural components in membranes (Hokin and Hokin, 1960). It was found that a secretory cell noticeably increased the formation of membrane phospholipids while active. In fact during its secretory activity there was a demonstrable increase in all the phospholipids with the exception of phosphatidylcholine. Hokin and Hokin also demonstrated that the salt glands in a resting stage contain a high concentration of phosphatidylinositol, which is reduced to diglycerides during secretion. Intracellular adenosine triphosphate donates a phosphate group to the diglyceride, producing phosphatidic acid. When the stimulus is removed the formation of phosphatidic acid ceases, whereupon it is reunited with free inositol to form phosphatidylinositol. Hokin has ventured to suggest that this mechanism plays a significant part in membrane circulation during the active periods but the precise role is still a matter of conjecture.

Those lipids encountered in membranes exhibit a wide variation in fatty-acid composition (see Table 4.2). Recent work suggests that membrane lipids associate with the structural proteins in such a way that the hydrophilic moieties at the membrane surface are determined by the internal protein which binds the hydrocarbon chains together (Figure 4.2).

In Chapter 3, page 85 it was inferred that the phospholipids were strongly hydrophobic hydrocarbon chains, possessing terminal polar groups capable of forming hydrogen bonds, and therefore hydrophilic. Since membranes are known to be a manifestation of specific spatial lipid—protein interaction it will be appreciated why the use of myelin figures and monolayer films is of extreme interest — since such studies of phospholipid in an aqueous system provide strong evidence that membranes are essentially bimolecular leaflet organizations. For instance if a phospholipid is placed in contact with a

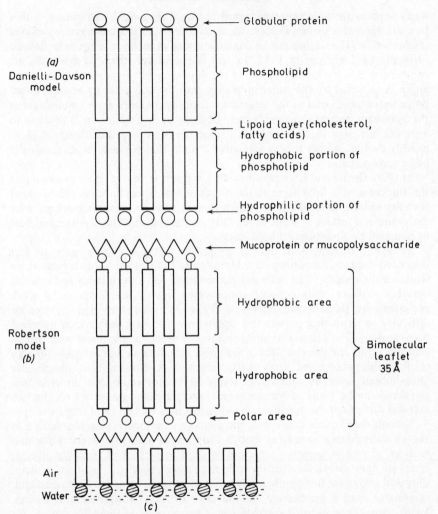

Figure 4.2. Biological membranes. (a) *The Danielli-Davson concept is based on phospholipid extraction studies from erythrocytes and myelin sheaths. Hence this model as opposed to the Robertson model, is based on physico-chemical criteria and not on electron microscopic studies.* (b) *The Robertson model shows the phospholipids in two distinct layers. It is proposed that the resident chains interdigitate forming one internal layer of phospholipid, by virtue of the fact that the overall length is actually 35Å, and evidence seems to suggest that the length of one chain is in the region of 28Å. The outer electron-dense layer is seen to be thicker than the internal, which is considered to be due to hydration of the hydrophilic carbohydrate side chains present in the outer part of the membrane.* (c) *Illustrates a lipoid film at an air-water interface (see text)*

water surface the molecules will spread out until they reach equilibrium – that is, until the entire surface is covered, constituting what is known as a monolayer (*Figure 4.2*). The technique of manufacturing monolayer films was devised originally by Langmuir in 1917. In such a monolayer, the lipid molecules are regimented side by side by virtue of the dipolarity of the molecules; the polar group is attracted by the water molecules, and the non-polar hydrocarbons are projected perpendicular to the interface – forming an unorientated monolayer at the boundary between air and water. Utilizing such a system it is possible to compute the area occupied by each molecule of, say, a monolayer of phosphatidylcholine, which is approximately double the cross-sectional area of the hydrocarbon chain.

In 1926, Gorter and Grendel utilized a Langmuir trough in order to determine the thickness of a lipid layer in the membrane of an erythrocyte. They found that the amount of lipid extract was sufficient to form a continuous layer over the entire cell surface – from which they concluded that the erythrocyte must be bounded by a bimolecular lipid layer.

In 1935, Danielli and Davson proposed that membranes were in fact organized laminas, consisting of a bimolecular lipid centre – as suggested by Gorter and Grendel – but with protein absorbed on to the interior and exterior complex surfaces. This concept had evolved out of the fundamental work performed by Danielli and Harvey in 1934–35, when they had experienced difficulty in extracting protein-free lipids from fish eggs with which they had been working. The classical Danielli membrane model which many workers still consider fulfils the prerequisites of the most recent data made available through electron microscope and x-ray diffraction studies, consists of a bimolecular phospholipid layer in which the non-polar hydrocarbons are directed perpendicular to the plane of the membrane; and the polar groups are on the two external surfaces of the membrane, sandwiched by a protein layer (*Figure 4.2*).

Though the Danielli concept of the membrane is widely accepted there is as yet no real evidence to suggest that it could not be otherwise. X-ray diffraction analyses of myelin suggest a structure believed to be formed from bimolecular layers of lipid orientated radially with alternating concentric layers of protein. Chemical analysis of this membrane type has shown it to comprise a phospholipid, cholesterol, and a cerebroside, in the ratio of 2:2:1 (Engstron and Finean, 1958). Also the work of Robertson seems to corroborate Danielli's theory, for hydrated monolayers of phospholipids form bimolecular layers which can be studied by both x-ray diffraction and electron microscopy. Their resemblance to biological membranes is striking – and furthermore they appear to be thermodynamically the most stable form of phospholipid–water complex.

Recently evidence has been published supporting the reality of lipid bilayers in plasma membranes using the freeze-etch methodology. Here tissues are rapidly frozen, then fractured and finally shadowed with platinum-carbon. The molecules which lie on the external surfaces of the membrane are revealed in bold relief. Evidence has been produced, to suggest, that hydrophobic lipid–lipid bonds along the central plane of a lipid bilayer would be the weakest point of the membrane in the frozen aqueous system. The exposed cleaved surfaces are lipid hydrocarbon on one side and hydrophobic protein on the other.

However it has been suggested that this Danielli bilayer–sandwich concept may not represent precisely the universal arrangement as postulated by Robertson, but may be regionally modified such that: (1) protein strands may penetrate from the membrane surface into its interior; or (2) some membranes may contain globular micelle pits interspersed within a protein lattice-work. Similarly some workers believe that membranes are composed of sub-units of lipoprotein associated with globular protein, resulting in a complex in which the hydrophilic groups of the protein and lipid are exposed to the aqueous environment, with the hydrophobic groups orientated towards the interior of the molecule.

Thus the question of internal organization of sub-units within the membrane structure is undecided since present data is insufficient to establish unequivocally a generalized conclusion. Suggested membrane structures are diagrammatically represented, in *Figure 4.2*.

CYTOSKELETAL MEMBRANES: STRUCTURE AND FUNCTION

Nuclear Membrane

The interphase nucleus of a cell is that area limited by a double membranous envelope approximately 200 Å thick which is known as the perinuclear cisternae. This membrane serves as the line of demarcation between the nuclear and cytoplasmic regions of the cell, which are known to be specifically different chemical and physical structures. The nuclear membrane is that structure via which genetical information – utilized in protein synthesis, for example – has to pass.

Electron microscopy has revealed that at regular intervals the inner and outer membranes of the nuclear envelope become fused forming nuclear pores or orifices. Studies on amphibian oocyte nuclei have shown that these nuclear pores are approximately 400 Å in diameter, and are disposed at intervals of 1 000 Å around the nuclear perimeter (Callan and Tomlin, 1950). Because of their apparent uniformity, and the fact that these annuli have been demonstrated in a variety of cells, they are no longer considered to be an artefact. The outer surfaces of the nuclear envelope possess ribosome particles which are adherent to its surface; this seems to suggest that the nuclear membrane is a derivative of the endoplasmic reticulum. It has been found that the nuclear membrane also possesses a significant electrical resistance, which is indicative of the presence of a diffusion barrier. Recent electron microscope studies of these nuclear pores have revealed that in some cell species they can become 'plugged'. Hence, correlation of this finding with the known electrochemical properties of the nuclear membrane, would seem to suggest that these pores are not freely communicating orifices. In fact it would appear that the nuclear membranes manifest a selective mechanism which is capable of diverting the random passage of, say, chromosome nucleoprotein from the nuclear region during the interphase period – yet at the same time permits a free interchange of ions between the nucleus and the endoplasmic reticulum.

The endoplasmic reticulum is an intricate labyrinth of membranous passageways the organization of which is in a constant state of flux. The reticular elements of the cytoplasm will vary considerably with the cell species, and they are divided into two general types – the granular (rough) and the agranular (smooth). Membranes of the latter kind are often associated with the Golgi apparatus. The former type of membranes are said to be rough when encountered in association with ribosomes which are arranged on the exterior surfaces of the reticulum membranes; in the smooth or agranular membranes, on the other hand, the ribosomes are either absent or present in relatively small numbers by comparison.

In cells actively involved in protein synthesis the granular endoplasmic reticulum is highly developed, sometimes consisting of large cisternae covered with ribosome particles. Examples of such reticulum membranes are to be found in the cells of the liver and pancreas. It is believed that the endoplasmic reticulum is a specific cytoplasmic expression of precise metabolic activity, the architecture of which is governed by laws as yet unknown. It has been postulated that the endoplasmic reticulum behaves rather as does a circulatory system, in motivating the import and export of the various substances disseminated throughout the intracellular matrix (Bennett, 1956).

The endoplasmic reticulum separates an extracisternal phase from an intracisternal phase: the cisternae are believed to perform transport functions principally, for it is assumed that structural protein units and other metabolic units are conveyed through these passageways to specific reception sites where they are utilized. They pass through the cisternae membrane wall of the endoplasmic reticulum to specific extracisternal sites such as a ribosome or mitochondria. In addition to serving as a transit route, the endoplasmic reticulum is known to be continuous with the plasma membrane so that by virtue of its structure and this association with the plasmalemma there is an increase in the area of the interal surface of the total structure. There is also evidence to support the involvement of the endoplasmic reticulum in the rapid conduction of nerve impulse in muscle. However, the precise nature and function of the endoplasmic reticulum has still to be resolved.

Golgi Complex and Dictyosomes

The Golgi apparatus or complex was for decades considered to be a histological artefact, but electron microscopy has revealed it to be otherwise. This cytoplasmic complex consists fundamentally of multiple flattened agranular sacs. Many workers consider the Golgi region to be a modified extension of the granular endoplasmic reticulum. Generally the appearance of the complex is of flattened cisternae which are arranged parallel to one another, the distal edges having broadened into spherical cavities surrounded by clusters of dense, small Golgi vesicles. In plant cells this Golgi complex is scattered throughout the cytoplasm in bodies called dictyosomes which morphologically are almost exclusively of the flattened type and surrounded by satellite vesicle clusters. In mammalian cells the Golgi complex is of a similar pattern. The thickness of the membranes is somewhere in the region of 60Å with an intercisternal space dimension varying between 50 and 200Å. At present very little is known about the role of the Golgi apparatus, and its association with the granular endoplasmic reticulum is conjectural. It has been suggested that the complex acts as a site of concentration for products elaborated elsewhere in a secretory cell. For instance, in the exocrine pancreatic cell, synthesized zymogen granules are believed to migrate from the endoplasmic reticulum to the Golgi apparatus where they condense, prior to their secretion from the cell apex, by means of a process considered to be the reverse of pinocytosis. Present chemical analyses of the Golgi complex membrane using centrifugation techniques suggest that there is very little, if any, enzymatic activity in their membranes.

Ribosomes

The ribosomes are ubiquitous structures of all living cells, being sub-microscopic particles composed of protein and ribonucleic acid. The number of ribosomes per cell is directly related to the ribonucleic acid content of that particular cell.

Much of our present knowledge with regard to the ribosome has evolved from ultracentrifugation studies. From centrifugation, the sedimentation coefficient can be determined directly, this varying for ribosomes between 70 and 80S, and their molecular weight has been calculated as $3-4.5 \times 10^6$. Generally, a ribosomal sub-unit is a sphere with flattened poles (an oblate spheroid). Two ribosome particles combined together form a dimer, and the association of two or more dimers results in a polyribosome or polysome. Ribosomal aggregates are also referred to as ergosomes since they are the active participants in protein synthesis. (*See* Protein Synthesis on page 205 for further discussion of this.)

Mictochondria and Choloroplast

The mitochondrion is to be found in all aerobic cells of animals and plants. The size and shape is variable, but usually the dimensions are approximately $5 \times 0.5 - 1\mu$ in diameter. The numbers found inhabiting a cell, again, will vary from 20 to 5×10^5, depending on the organism concerned. It is estimated that in a normal liver cell there are some 2 000 mitochondria, but this figure drops during regeneration, and also in tissue where there is cancerous involvement.

The ultrastructural morphology of a mitochondrion is that of an elongated egg. The mitochondrion is enveloped by an outer limiting membrane approximately 60 Å thick. Beyond this membrane is an aqueous phase of some 80Å, and then an inner membrane which is also some 60 Å thick. This latter membrane is seen to be invaginated, forming internal partitions known as the cristae mitochondriales, which project into the mitochondrial matrix.

Electron microscopy of these mitochondrial membranes has shown them to consist of two layers of high electron density, sandwiching a lighter middle layer. This arrangement complements the Robertson structure which was mentioned in detail on page 99. From this evidence it has been assumed that the mitochondrial membranes are formed from two lipid layers the non-polar groups of which are orientated in the centre — which would represent the light 'filling'

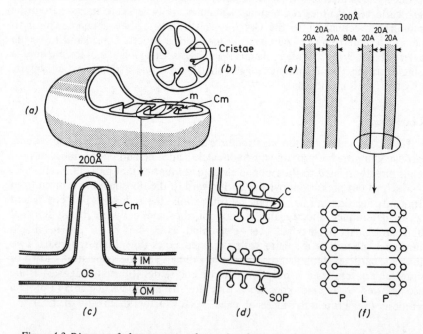

Figure 4.3. Diagram of ultrastructure of mitochondrion. (a) General three-dimensional diagram of a mitochondrion showing: (m) matrix and invaginated membranes forming the cristae mitochondriales (Cm). (b) Shows mitochondrion in cross-section. (c) Indicates a higher magnification of the cristae mitochondriales (Cm); IM is the inner membrane; OS the outer space; OM the outer membrane. (d) Shows the cristae (c) with the sites of oxidation phosphorylation (SOP). (e) Is a higher magnification of (c) indicating the inner and outer membrane dimensions. (f) Molecular diagram showing characteristic protein–lipid–protein sandwich — P is protein and L is lipid

of the membrane sandwich; the two dark outer regions represent the protein or crust. The ultrastructure of a mitochondrion is shown in *Figure 4.3* together with a diagrammatic impression of a single cristae and a molecular interpretation of the electron microscope image.

The inner membrane of the mitochondrion appears to be smooth, but if examined under an electron microscope, following staining with phosphotungsten, it is seen to be covered with spherical sub-units called elementary or sub-mitochondrial particles. The particles consist of a stalk and terminal headpiece some 80–100 Å in diameter, these being evenly distributed along the inter-mitochondrial membrane surfaces. Their significance, and the role played by the mitochondrion unit in electron transportation and oxidative phosphorylation will be discussed in Chapter 5 which deals with Bioenergetics.

By virtue of the continuity of the mitochondrial membranes with those of the plasmalemma, endoplasmic reticulum and nuclear membrane, it has been suggested that the mitochondria arise from specific sites within the cytoskeletal membrane system. However, exactly what mechanism, genetical or otherwise, initiates and controls mitochondrogenesis is still a matter of conjecture, but it has been shown that the half-life of an average mitochondrion is in the region of 5–10 days, which would seem to indicate that their synthesis is a continuous one.

The plastids are cytoplasmic organoids which are widely distributed in the plant kingdom, and are also to be found in a few autotrophic protozoans. Under the term plastid may be included chloroplasts, leucoplasts – if colourless– and chromoplasts – if coloured. It is commonly assumed that all these bodies arise from smaller bodies called proplasts. The chlorophyll-containing plants – that is those containing chloroplasts – have been studied in some detail but this is hardly surprising since they are the most common, and of great economic and biological importance. This, since the animal kingdom is absolutely dependent for its continuing survival upon the products of photosynthesis. In almost all photosynthetic organisms, photosynthesis occurs in the chloroplasts which contain chlorophyll. Hence the chloroplast is intimately related to the metabolic processes of plant cells, as are the mitochondria to those of the animal cell – for the primary function of both is the generation of energy which can be utilized in cellular metabolism. In the chloroplast the external energy source is light, and in the mitochondrion it is in the form of oxidizable organic substrates which can enter the cell. Both types of energy-yielding processes, and their connection with the mitochondrion and chloroplast respectively will be discussed later in the text (*see* page 110). Only the general structure of the chloroplast will be considered here.

As already implied the chloroplasts are characterized by the presence of the pigment chlorophyll, which consists of a hydrophilic head composed of a tetrapyrrole ring situated about a magnesium atom, and a long hydrophobic chain (*Figure 4.4*); collectively this is known as a metalloporphyrin complex. The structural relationships of chlorophyll, cytochrome c, and phycobilin, to

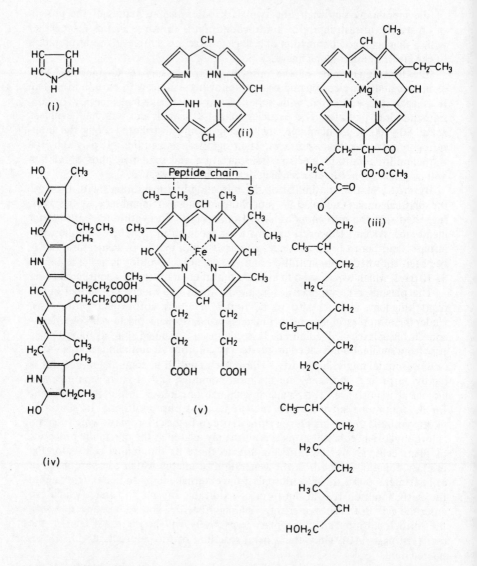

Figure 4.4. Structural relationship of chlorophyll and others to porphin. (i) The basic pyrrole unit; (ii) porphin; (iii) chlorophyll showing long lipid tail; (iv) phycobilin (a protein-conjugated pigment found in blue-green algae); (v) the respiratory pigment cytochrome c, which forms part of the electron transfer system – indicating part of peptide chain to which cytochrome is conjugated

(a)

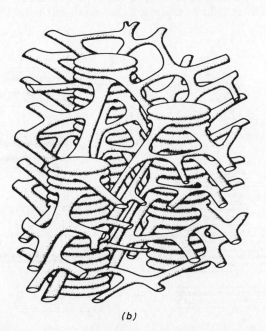

(b)

Figure 4.5. (a) *Model of chloroplast, showing the inner structure, with the grana disposed in stacks perpendicular to the surface.* (b) *Drawing of chloroplast ultrastructure indicating the interconnecting network of anastomosing tubules that unite the membranous compartments of the grana (see Figure 4.6) (from Weier and colleagues, 1963; reproduced by courtesy of the authors and Academic Press)*

porphyrin are indicated in *Figure 4.4*. Chemical analysis of chloroplasts shows that approximately 55 per cent of the dry weight is protein which is intimately bound to some 30 per cent of lipids, so forming a lipoprotein complex. The lipid fraction consists of approximately 50 per cent fats, 20 per cent sterols, 15 per cent waxes and 5 per cent phospholipids. Chlorophyll is also seen to constitute approximately 9 per cent of the dry weight of which 75 per cent is in the form of chlorophyll c and the remainder of chlorophyll b. Other pigments belonging to the carotenoid group are also found in the proportion of approximately 5 per cent, consisting of 75 per cent xanthophyll and 25 per cent carotenes. The size, shape and distribution of the chloroplasts in a plant cell will vary according to the species under consideration. Chloroplasts in the higher plants tend to exhibit a certain degree of structural conservation. In the main such plants will contain a large number of discoid, spheroid or ovoid chloroplasts approximately 5 μ in diameter, and up to 3 μ in thickness. Their distribution in the plant cell cytoplasm is dependent largely on conditions in the external environment; of these, light is the primary consideration. Under the high power of

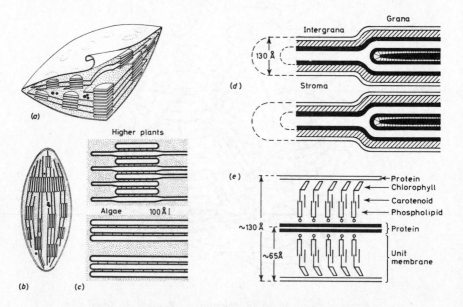

Figure 4.6. Fundamentals of chloroplast structure and development: (a) *schematic three-dimensional structure of chloroplast of a higher plant;* (b) *cross-section through (a);* (c) *enlargement of (b) showing details of lamellar structure and a comparison between that of higher plants and algae;* (d) *schematic interpretation of the lamellation in the chloroplast;* (e) *molecular diagram of (d) in terms of protein, lipid and chlorophyll (from Mahler and Cordes, 1968; reproduced by courtesy of Professor von Wettstein and Harper & Row)*

an ordinary light microscope they show the presence of a disc-shaped grana which is embedded in the stroma. The membranes of the grana appear to be largely lipoprotein in nature, whereas the stroma is protein. With the electron microscope the chloroplast is seen to consist of parallel lamellae of lipoprotein, such that the grana are composed of disc-shaped lamellae, which can be likened to a pile of coins (*Figure 4.5*). Each grana is approximately $0.3-0.4 \, \mu$ in diameter, there being from 10–100 per laminae. The ultrastructure of the laminae is somewhat analogous to that of myelin. A schematic, and somewhat speculative picture of chloroplast lamellation is outlined in *Figure 4.6*. In electron micrographs there are areas of increased lamellar density which are formed by the interpolation of lamellae. From this interpretation of the chloroplast, the intergrana material is found to be comparable in composition to the grana—hence, photosynthesis is considered to take place throughout the chloroplast, since chlorophyll is uniformly distributed throughout the lamellar structure. Electron microscopic observation suggests that the chlorophyll and lipid molecules are arranged in a double layer of some 50 Å thickness, each layer being coated with a layer of protein. It would seem that the role of the lipid layer is one of insulation, in the sense that, as a result of light absorption, separation of the initially formed reactive radicals is achieved; and they are not permitted to recombine haphazardly prior to performing useful chemical work (*see* Photosynthesis on page 147).

Lysosomes

Lysosomes and the lysosomal apparatus play an important functional role in the homoeostasis of a cell. Fundamentally the lysosomal apparatus (lysosomes and phagosomes etc) is composed of a variety of cytoplasmic vacuoles which are

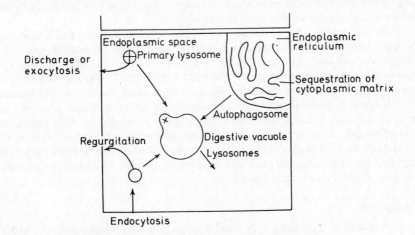

Figure 4.7. Stylized scheme of lysosomal apparatus

111

thought to arise from the endoplasmic reticulum and Golgi region in response to engulfed material. Lysosomes are characterized by the presence of various hydrolases and, in certain cell species, of bacteriostatic basic proteins. Primary lysosomes are those structures not yet involved in a digestive event, which discharge their contents either into the cell exterior or into a digestive vacuole or secondary lysosome which contains the substances of intracellular digestion. Extracellular substances are carried from the cell surface — phagosomes to the digestive vacuoles, where they are exposed to the digestive process with secondary lysosomes. The relationship between the lysosome apparatus and the nutritional and phagocytic activity of a cell is shown in *Figure 4.7.*

Phagocytosis and Pinocytosis

In addition to exhibiting mechanisms of passive diffusion and active transportation of ions and molecules across membranes — which will be discussed in the chapter dealing with bioenergetics — the uptake of material from the surrounding environment can occur by means of certain cytoplasmic convulsions, as represented by the process of phagocytosis and pinocytosis.

Phagocytosis

As performed by such primitive cells as the protozoon amoeba, phagocytosis or engulfment represents a normal feeding mechanism. In the higher animals, and particularly in the mammals, phagocytosis is a highly developed activity, as exemplified by the granular leucocytes, and the cells of the reticulo-endothelial system. These cells are able to ingest bacteria and other cell debris in the same way as the amoeba feeds, except that here the action has been modified into the total somatic organization as a defence mechanism.

Cells performing such a function are known as macrophages. It is interesting to note here that certain bacteria which are pathogenic have developed anti-phagocytic capsules, this contributing to their virulence — for example, the capsule of *Streptococcus pneumoniae.* Similarly some bacteria and also viruses have developed a biochemical character which enables them to survive the trauma of engulfment, and subsequently to multiply in an intracellular position; an example of such a bacterium would be *Neisseria meningitidis.* This ability to specialize and adapt to a foreign micro-environment has not yet been explained. As already implied phagocytosis is simply a process whereby a foreign particle is engulfed by a protoplasmic mass and invaginated, thereby forming a cytoplasmic vacuole which in due time is digested.

Pinocytosis

Pinocytosis is a process very similar in action to phagocytosis, by which substances such as fats and proteins can be taken up at the cytoplasmic membrane

112

interface, with the subsequent formation of an intracellular pinocytic channel (*Figure 4.8*). As the engulfed substance flows inward, fragmentation of the intra-terminal end of the channel occurs with the formation of micro-vesicles which can be slowly absorbed and utilized as required in the cytoplasm. The secretory activity of exocrine glands is considered to constitute the reverse of the processes briefly described here.

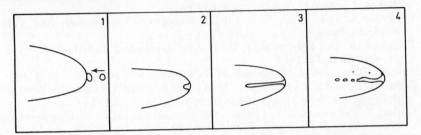

Figure 4.8. Diagrammatic series indicating the fate of an engulfed liquid – for example, a fat globule. (1) and (2) show initial contact; (3) shows the formation of a tubular channel; and (4) its subsequent fragmentation

Of interest in this context is the fate of foreign antigenic material in the animal body, with respect to the eventual interaction of residual antigen with antibody subsequently formed. For instance when an antigen is introduced into a system it is phagocytosized generally, a process occurring more readily if it is associated with particular matter – for example, pneumococcal polysaccharide complexed with bacterial protein. Within the mammalian reticulo-endothelial system there are various kinds of cells exhibiting such a capacity, these either being fixed in tissue complexes or mobile within the general circulatory system. Some cells on the other hand possess the ability to take up soluble antigen by microphagocytosis – that is by the process of pinocytosis (Rowley, 1962). The fate of an antigen, and the role of the macrophage in antibody synthesis will be discussed in more detail in Part V which deals with the maintenance of somatic integrity.

BIBLIOGRAPHY AND REFERENCES

Bennett, H. S. (1956). 'The Concepts of Membrane Flow and Membrane Vesiculation as Mechanisms for Active Transport and Ion Pumping.' *J. biophys. biochem. cytol.*, Suppl. 2., 99.

Bourne, G. (1951). 'The Mitochondria and Golgi Apparatus.' In *Cytology and Cell Physiology*. Ed. by G. Brown, London; Oxford University Press.

Brachet, J. and Mirsky, A. E. (Eds.) (1961). *The Cell*. New York; Academic Press.

Callan, H. G. and Tomlin, S. G. (1950). 'Experimental Studies on Amphibian Oocyte Nuclei. I. Investigation of the Structure of the Nuclear Membrane by Means of the Electron Microscope.' *Proc. R. Soc.*, B137, 367.

Chapman, D. (1968). *Biological Membranes*. London; Academic Press.

Danielli, J. F. and Davson, H. (1935). 'A Contribution to the Theory of Permeability of Thin Films.' *J. cell comp. Physiol.*, 5, 495.

Davson, H. and Danielli, F. J. (1952). *The Permeability of Natural Membranes*, 2nd ed. London; Cambridge University Press.

Engstron, A. and Finean, J. B. (1958). *Biological Ultrastructure*. New York; Academic Press.

Finean, J. B. and Robertson, J. D. (1954). 'Lipids and Structure of Myelin.' *Br. med. Bull.*, 14, 267.

Fawcett, D. W. (1961). 'Membranes of the Cytoplasm.' *Lab. Invest.*, 10, 1162.

Gunsalus, I. C. and Stainer, R. Y. (1964). *The Bacteria: A Treatise on Structure and Function*. New York; Academic Press.

Gorter, E. and Grendel, F. (1926). 'On Bimolecular Layers of Lipoids on the Chromocytes of the Blood.' *J. exp. Med.*, 41, 439.

Granick, S. (1961). 'The Chloroplasts: Inheritance, Structure and Function.' In *The Cell*, vol 2, page 489. Ed. by J. Brachet and A. E. Mirsky. New York; Academic Press.

Ham, A. W. (1965). *Histology*, 5th ed. Philadelphia, Pa; Lippincott.

Hokin, L. E. and Hokin, M. R. (1960). 'Studies on the Carrier Function of Phosphatidic Acid in Sodium Transport. I. The Turnover of Phosphatidic Acid and Phospho-inositide in the Avian Salt Gland on Stimulation of Secretion.' *J. gen. Physiol*, 44, 61.

— — (1965). 'Chemistry of the Cell Membrane.' *Scient. Am.*, 213, 78.

Langmuir (1917) *see* Ponder.

Lehninger, A. L. (1964). *The Mitochondrion*. New York; Benjamin.

Mahler, H. R. and Cordes, E. U. (1968). *Basic Biological Chemistry*. London; Harper and Row.

Miles, A. A. and Pirie, N. W. (1950). *The Nature of the Bacterial Surface*. Springfield, Ill.; Thomas.

Muhlethaler, K. (1961). 'Plant Cell Walls.' In *The Cell*, vol 2, page 85. Ed. by J. Brachet and A. E. Mirsky. New York; Academic Press.

Novikoff, A. B. (1961). 'Mitochondria.' In *The Cell*, vol 2, page 299. Ed. By J. Brachet and A. E. Mirsky, New York; Academic Press.

Palade, G. E. (1964). 'The Organisation of Living Matter.' *Proc. natn. Acad. Sci. USA.*, 52, 613.

Ponder, E. (1961). 'The Cell membrane and its Properties.' In *The Cell*, vol 2. Ed. by. J. Brachet and A. E. Mirsky, New York; Academic Press.

Robertson, J. D. (1961). 'Cell Membranes and the Origin of Mitochondria.' In *Regional Neurochemistry*, page 497. Ed. by S. S. Kety and J. Elkes. Oxford; Pergamon Press.

Rowley, D. (1962). 'Phagocytosis.' In *Recent Advances in Immunology*, page 241. Ed. by W. H. Taliaferro and J. H. Humphrey. New York; Academic Press.

Rustad, R. C. (1961). 'Pinocytosis.' *Scient. Am.*, 204, 121.

BIBLIOGRAPHY AND REFERENCES

Strominger, J. L. and Tipper, D. J. (1965). *Am. J. Med.*, **39**, 708.
Ussing, H. H. (1957). 'General Principles and Theories of Membrane Transport.'
   In *Metabolic Aspects Across Cell Membranes*. Ed. By R. R. Murphy.
   Madison, Wis.; University of Wisconsin Press.
− (1960). 'Physiology of the Cell Membrane.' *J. gen. Physiol.*, **43**, 5. Suppl. 135.
Waddington, C. H. (1959). *Biological Organisation: Cellular and Subcellular.*
   Oxford; Pergamon Press.
Weier, T. E., Stocking, C. R., Thomson, W. W. and Prever, H. (1963). *J.
   Ultrastruct. Res.*, **8**, 122.

Part III
The Molecular Basis of Cellular
Energy Transformations

5 — Bioenergetics

(Basis of bioenergetics − Adenosine triphosphate: structure and function − Oxidation and reduction − Anaerobic respiration − Aerobic respiration and the mitochondrion − Photosynthesis − Muscular contraction − Active transport − Action potentials)

THE BASIS OF BIOENERGETICS

In Chapter 2 it was briefly shown how the form and function of cellular organization was simply an expression of complex energy transformation systems which were intimately linked with a mainstream of energy flowing through the biological world and which could be explained accordingly by the principles of thermodynamics.

Energy is often defined as the capacity to do work, or as the product of a particular force (F) acting through a given distance (D): that is, $W = F \times D$, where W is the resultant mechanical energy. But energy states may be manifested in a variety of forms apart from mechanical work, examples being kinetic, electrical and potential energy.

Internal Energy

Every chemical system has internal energy (E), which is a function, among other things, of the temperature and the nature of the substance involved, and since almost all chemical and physical reactions are accompanied by a loss from, or gain to, the surrounding environment it is the change in internal energy which is of fundamental interest. When there is a loss of heat to the surrounding environment as a result of an event, the reaction is called exothermic. Alternatively, when it occurs with the absorption of heat from the environment the reaction is said to be endothermic. Consider that heat required to melt substance x at its melting point: here it is the change in internal energy which is of prime importance, and this can be designated function ΔE. Hence if E_2 represents the internal energy status of x in the liquid state, and E_1 the internal energy status in the solid state then:

$$\Delta E = E_2 - E_1 \ldots (1)$$

119

According to the first law of thermodynamics, energy is neither created nor destroyed – that is the total energy of the universe always remains constant. Expressed mathematically the first law is:

$$\Delta E = q - w \ldots (2)$$

where q represents the heat change, and w the amount of work performed by that system. By virtue of the principle of the conservation of energy, any heat additions to a system (q) are manifested by a reciprocal change in its internal energy ΔE, which represents the amount of work done by that system. Thus:

$$q = \Delta E + w \ldots (3)$$

Returning to equation (1): ΔE is a function of state, and thus independent of the path taken in going from the initial to the final state. Since thermodynamic functions of state give a description of the thermodynamic properties of a system in a particular state, in no way is the value of E affected. But, in the change, the value of ΔE may be seen to be negative or positive depending on whether E_2 is greater or less than E_1 (equation 1).

The gain or loss of heat in a reaction may be determined using an apparatus called a calorimeter, which provides a simple means of measuring directly the amount of energy released as a result of a reaction. When a chemical reaction occurs at a constant pressure, the thermodynamic function ΔH is used to express what is termed the enthalpy or heat content of the system. Thus by substituting ΔH for q in equation (2) we get:

$$\Delta H = \Delta E + w \ldots (4)$$

Using the calorimeter, the heat of the reaction at a constant volume and pressure is determined – here no work is produced. Hence, equation (4) becomes:

$$\Delta H = \Delta E \ldots (5)$$

Exothermic and Endothermic Reactions

Reactions can be classified either as exothermic ($- \Delta H$), or endothermic ($+ \Delta H$), depending on whether heat is evolved or absorbed. Take for instance the following reactions:

(A) $C + O_2 \rightarrow CO_2 \quad \Delta H = -94 \cdot 05$ kcal
(B) $2C + 2H_2 \rightarrow C_2 H_4 \quad \Delta H = +12 \cdot 5$ kcal

In (A) the ΔH of the reaction is a negative value so that the reaction is exothermic; in (B) the ΔH value is positive indicating that the reaction is endothermic. The use of the convention ΔH indicates that the reaction occurred

120

at a constant pressure. Equation (A) is an example of the heat of combustion which can be defined as the heat surrendered to the surrounding environment in terms of calories when one gramme-molecular weight of a substance is combusted completely at the expense of molecular oxygen. It will be seen that such a reaction proceeds to completion with a definite heat of reaction which is quantitatively related to the number of reacting molecules. Originally a calorie was taken as that unit of energy required to raise the temperature of one gramme of water from 15 to $16°C$. One gramme-calorie of heat energy is seen to be equivalent to 4.184×10^7 ergs — since one erg is defined as the work performance when a force of 1 dyne (that force imparting to one gramme an acceleration of 1 centimetre per second) acts through a displacement of one centimetre. Since an erg is an inconvenient unit, the term absolute Joule is employed; 1 Joule $\equiv$ 10^7 ergs and 4.184×10^7 ergs becomes 4.184 Joules. From which heat can be equated with mechanical work. The efficiency of mechanical heat engines to perform work is dependent on the existence of a considerable temperature differential, whereas in the living cell there is no such restriction since they are essentially isothermal. In relation to the second law of thermodynamics this fact would seem to pose a paradox.

Generally speaking the second law of thermodynamics is a logical extension of the first law. The first law was seen to place no restriction on the transformation of energy, since it merely defined the conservation of energy, when energy changes occurred. The second law on the other hand clearly relates information regarding the conditions under which heat may be converted into other forms of energy, and the quantitative limitations of such a conversion. Thus, the second law states that all systems in the universe tend to move to a maximum state of randomness, expressed by the symbol (S) — which is termed entropy. Combining the two laws, therefore, we can say that the energy of the universe is constant, but the entropy approaches a maximum. To elaborate on this, let us consider two bodies, one hot and the other cold. Heat from the hot body will flow spontaneously to the cold body, and the temperature of the initially cold body will rise until both bodies reach a temperature equilibrium. All systems according to the second law will endeavour to reach this state of equilibrium at which all measurable physical and chemical parameters of state are uniform throughout: that is, at which they have become randomized to the maximum. From this example of the two bodies at different temperatures in which maximum randomization has occurred, it is apparent that a spontaneous reversion of this state will not occur of its own accord. First then, the concept of entropy is seen to be a measure of randomness, which is energy unavailable to do work; secondly, all physical and chemical processes will proceed towards the attainment of this state.

The third law of thermodynamics states that the entropy of any element or compound in its normal crystalline state at absolute zero temperature is zero. This concept assumes that at a temperature of absolute zero there will be no thermal motion, and that the atomic structure of the crystal in question will be in a state of maximum order, which is the complete opposite of entropy. Hence assuming that entropy will be zero at absolute zero of temperature, it is possible to calculate the absolute entropy of a substance at any temperature. The value for the absolute entropy of many substances has been determined at a temperature of 25°C. These entropy values are expressed in terms of entropy units ($S°$) — that is, calories per degree per mol. Table 5.1 lists some representative thermodynamic values.

TABLE 5.1

Some Representative Thermodynamic Values

	$\Delta H°$ kcal/mol	$\Delta S°$ cal/deg/mol	$\Delta G°$ kcal/mol
H_2O	− 68·317	16·72	− 54·36
Acetic acid	−116·4	−	− 93·8
Oxidation of glucose	−673	−	−686
Hydrolysis of glucose-6-phosphate	− 3	−	− 3·3

The calculation of $\Delta S°$ is the same as for ΔH:

$$\text{for reaction } A_a + B_b \rightleftharpoons C_c + D_d$$

$$\Delta S° \text{ will be } \Sigma S° \text{ (products)} - \Sigma S° \text{ (reactants)}$$
$$\text{hence} = CS°_c + DS°_a - AS°_a - BS°_b$$

Thus a decrease in entropy within the universe must be accompanied by a reciprocating increase in some other part of it, in order to satisfy the first law. The second law of thermodynamics shows that spontaneous chemical or physical changes have direction, with a tendency to seek maximum entropy. Consider now the chemical reaction between zinc metal and acid:

$$Zn + 2H^+ \rightarrow Zn^{2+} + H_2$$

There is a spontaneous reaction in which hydrogen is released together with a quantity of heat energy which continues until a state of equilibrium is reached — that is, the evolved heat is absorbed by the system, and thereby the system and its surroundings reach a state of maximum entropy. In every spontaneous change there is made available a quantity of energy that can be utilized for useful work, this being termed free energy. In this case the available heat energy evolved was allowed to dissipate, but if, on the other hand, the same reaction was allowed to take place in a controlled system, such as an electrical cell, the free energy so produced can be retained in the form of electrical energy and utilized in useful work. The free energy content of a system is described by the thermodynamic function of state G, and is a measure of the potential energy of a system which can do work under isothermal conditions. But it is also a quantity that cannot be measured directly. In the example given above — that is, for the reaction between zinc and acid — the relation here between the free energy change and the electromotive force of the cell is given as:

$$\Delta G^\circ = -E^\circ n.F \ldots (6)$$

where G° represents the free energy produced in a reaction; E the standard electromotive force of the cell; n the number of electrons transferred and F the Faraday constant.

Considering the relationship of free energy changes (ΔG) to changes in enthalpy (ΔH), the total measurable energy change, and entropy (ΔS), the total measurable energy change at constant temperature (T) is:

$$\Delta G = \Delta H - T\Delta S \ldots (7)$$

Equation (7) implies that a decline in ΔG will be accompanied by a reciprocal increase in $T\Delta S$. The expression in equation 7 assists in an easier comprehension of the state of entropy since, when ΔS is negative, it implies that there is an increase in order. For if heat is lost from a system then the decline in ΔG will be greater than the gain in ΔS. Likewise with the absorption of heat, the decline in ΔG will be less than the gain in ΔS of the system.

The free energy changes in the reaction $A + B \rightleftharpoons C + D$, can be calculated thus:

$$\Delta G^\circ = \Sigma G \text{ (products)} - \Sigma G \text{ (reactants)}$$

$$\Delta G = \Delta G^\circ + RT \log_e [C][D]/[A][B]$$

Equation (8) provides a means for the experimental determination of the standard free energy change of reaction (ΔG°) from which the relation between it and the equilibrium constant can be shown. Hence for the reaction $A + B \rightleftharpoons C + D$:

$$K = [C][D]/[A][B]$$

$$\log_e K = (-\Delta G^\circ)/RT \text{ thus: } \Delta G^\circ = RT. \ln k \ldots (8)$$

where ΔG° is the standard free energy change; R is the gas constant; T is the Kelvin temperature and lnk is the natural logarithm of the equilibrium constant.

At equilibrium there is no change in free energy, $\Delta G = 0$, therefore:

$$\Delta G = \Delta G^\circ + RT. \, Log_e \, [C] \, [D] / [A] \, [B]$$
$$\text{hence } 0 = \Delta G^\circ + RT. \, Log_e \, [C] \, [D] / [A] \, [B]$$

changing logs to the base of 10,

$$0 = \Delta G^\circ + 2 \cdot 303. \, RT. \, Log \, K$$
$$\Delta G^\circ = -2 \cdot 303. \, RT. \, Log \, K$$

For example, the standard free energy change for the complete oxidation of glucose to carbon dioxide and water may be obtained thus:

$$\text{Glucose} + 6O_2 \rightleftharpoons 6CO_2 + 6H_2O$$
$$\Delta G^\circ = 6(56 \cdot 69 + 94 \cdot 26) - 219 \cdot 22 = -686 \cdot 48 \text{ kcal/mol}$$

Thus when ΔG is large and negative the reaction will go virtually to completion. If on the other hand ΔG is large and positive then K is low, which is one way of saying that the reaction will not proceed far in the direction of completion. It is important to point out here that ΔG indicates only the extent of the reaction, and not how far it will proceed. Reactions having a negative ΔG are said to be exergonic, and those with a positive ΔG are termed endergonic. In the case of an exergonic reaction it is implied that the reaction will proceed spontaneously, accompanied by a heat loss, and that work may be obtained from the system, giving rise to an increase in entropy. Endergonic reactions on the other hand, where ΔG is positive, require an external supply of energy to make them go – this resulting in an uptake of energy, and a subsequent decrease in entropy. As already stated ΔG fails to convey any information regarding the speed with which, or indeed the manner in which, a reaction will proceed. A chemical reaction is a process by which the identity of molecules becomes changed; this often requires special conditions in order that the atoms of the reactants may more readily be in close apposition, such that the orbitals of their electrons overlap, thereby precipitating the rupture or exchange, and subsequent formation of new molecular bonds. Hence, only as a result of additional energy put into the system will a reaction go spontaneously. This additional energy is referred to as the activation energy, and is one of the prime factors involved in the determination of the speed of a reaction.

Consider briefly the reaction between hydrogen and oxygen, which will combine thermodynamically to form water (*Figure 5.1 a*). Here $\Delta G'$ represents the quantity of activation energy required to bridge the energy barrier, and thus make the reaction go, and proceed to equilibrium. In fact for this reaction $\Delta G = -120\,000$ cal/mol of water. Actually the reaction will not proceed spontaneously at room temperature. The value of activation energy is always positive — that is, energy must be added to the system. In *Figure 5.1b* reactants A must acquire again activation energy E_1; B represents the activated state. E_2 is equal to the loss of activation energy on the formation of the products at C. According to *Figure 5.1b,* therefore, ΔG is given by the distance AC.

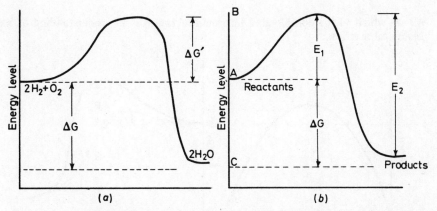

Figure 5.1. Graphs illustrating text: (a) *representation of the thermodynamic reaction between hydrogen and oxygen;* (b) *activation energy of reactants*

The Law of Mass Action states that the rate of chemical reaction at any given time is proportional to the molecular concentration of the reacting substances. Thus:

$$A + B \underset{v_2}{\overset{v_1}{\rightleftharpoons}} C + D$$

and therefore $-d[A]/dt = -d[B]/dt + d[C]/dt - d[D]/dt$

therefore $d[A]/dt.[A][B] = -d[A]/dt = k[A][B]$

from which it can be shown that the activating energy is related to the velocity constant, for:

$$d[Ink]/dt = \Delta G/RT$$
$$\therefore Ink = -\Delta G/RT + 1, \text{ since } k_1[A][B] = k_2[C][D]$$
$$\text{thus } \log k_2/k_1 = \Delta G/2 \cdot 3R.(1/T_1 - 1/T_2)$$

Temperature will greatly influence reaction velocity since molecular movement is increased as also is the chance of molecular collision between the reactants (*Figure 5.2a*). Standard enthalpy changes in a reaction can be calculated from the measurements of equilibrium constants as a function of temperature:

$$d\, Ink/dt = \Delta H^\circ/RT^2 \; ; \therefore Ink = \Delta H^\circ/RT + \text{constant}$$

125

A plot of log K against the reciprocal of the absolute temperature yields a straight line of slope $-\Delta H^{\circ}/2 \cdot 3R$, thereby evaluating ΔH°. The standard entropy change ΔS° can be determined from the relationship:

$$\Delta S^{\circ} = (\Delta H^{\circ} - \Delta G^{\circ})/T$$

All of which yields considerable information towards the understanding of a biological reaction.

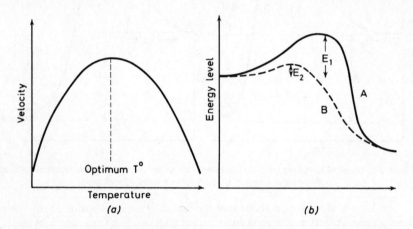

Figure 5.2. (a) *The effect of temperature on reaction velocity;* (b) *the effect of the presence of a catalyst in a system (see text)*

Similarly a catalyst will accelerate the rate of a reaction. A catalyst is a substance capable of increasing the rate of a reaction, and yet of remaining chemically unchanged at the end of the reaction, and of not having altered the position of chemical equilibrium. That is, the value of K does not change, since K is determined solely by ΔG°.

In *Figure 5.2b*, the presence of a catalyst in a system is illustrated, showing the effect of a reduction of the activation energy required to convert reactants into products. Line A indicates the normal activation energy barrier pathway, whose extent is E_1. Line B shows the reduction in the quantity of activation energy required as a result of the presence of a catalyst.

The thermodynamic principles considered briefly above apply to closed systems, by which is meant a system isolated from its surroundings, and one which does not exchange material with the external environment. Equilibrium is attained in such a system in which reactions balance each other, and entropy is always maintained if reactions are reversible or increased if they are non-reversible. Thus according to the second law of thermodynamics, it is seen that the entropy of a closed system tends to increase progressively. Unlike a heat engine, a living system appears to oppose the concept of the second law of thermodynamics (entropy), since cells are characterized by an increase in order and organization. This paradox is resolved by the concept that living cells and the biosphere constitute an open system. An open system is one in which matter and energy exchanges are continuously taking place between the components of the system and the external environment, such that there is a continuous flow of products through the entire system (*see* Chapter 2, page 40). Indeed the rate of formation of products within the cell is equally balanced by the rate of degradation, and their removal to the external environment — all of which maintains the system in what is considered to be a dynamic steady state.

ADENOSINE TRIPHOSPHATE:
ASPECTS OF ITS STRUCTURE AND FUNCTION

Following a redistribution of energy within a system the resultant energy manifested may be lost to the system in the form of heat, utilized in the performance of cellular work, or conserved by the formation of new chemical bonds which when broken will release a large quantity of energy. Such bonds are known as energy-rich bonds which serve as the principal source of energy required to drive the numerous endergonic processes within the cell. By far the most common example of this kind of energy-rich bond are the high energy phosphate bond groups which are probably best characterized by adenosine triphosphate (ATP):

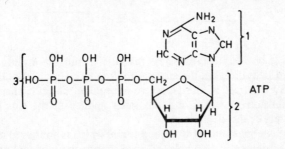

Structurally, as shown, ATP consists of three units: (1) a heterocyclic aromatic ring called adenosine to which is attached by glycosidic linkages (2) the five-carbon sugar D-ribose, to which is attached in turn, (3) the phosphate group in ester linkage at the 5′ position. Other examples of monucleotides found in the cell are adenosine diphosphate (ADP) and adenosine monophosphate (AMP).

Molecules such as those of ATP are characterized by a large negative free energy of hydrolysis, under physiological conditions. The hydrolysis of ATP yields a greater change in free energy than does that of, say, glucose-6-phosphate. Thus:

$$\text{ATP} \rightleftharpoons \text{ADP} + \textcircled{P} \text{ (at 25°C and pH 7·0)}$$
$$\Delta G = -7\ 000\ \text{cal/mol}$$

whereas:

$$\text{glucose-6-phosphate} + H_2O \rightleftharpoons \text{Glucose} + \textcircled{P} \text{ (at pH 7·0)}$$

$$\Delta G = -3\ 300\ \text{cal/mol}$$

As with ATP, the hydrolysis of ADP to AMP is strongly exergonic — in fact at pH 7·0 the $\Delta G = -6\ 500$ cal/mol. By contrast the ΔG of hydrolysis of AMP is appreciably less, and in fact it is only equivalent to the yield of the hydrolysis of a simple phosphate ester. Hence from the examples of hydrolysis of ATP and glucose-6-phosphate it is apparent that an energy-rich compound may be defined as one whose reaction with a substance commonly present in the environment is accompanied by a large negative free energy change at a physiological pH (Jencks, 1962).

The reason for ATP having a large $-\Delta G$ of hydrolysis is attributed to the salient feature of the highly charged polyphosphate structure. Primarily then the molecule inherently manifests an electrostatic repulsion between the positive charges of the phosphorus atom, which evolves from the attraction of electrons to the oxygen double bond:

Hence there is an electrostatic stress resident within the structure. When the terminal bond is hydrolysed the energy required to maintain the above structure is released. By contrast in the glucose-6-phosphate molecule there are no such forces of repulsion at work, and the same applies in the simple low-energy phosphate esters of the alcohols.

In ATP the energy status of this bond is one which is conferred upon it by the molecular environment as a whole, and not simply an individual property of the phosphate bond.

In the cell the ATP and ADP molecules are stabilized by forming a complex with Mg^{++} or Ca^{++}. Illustrated in *Figure 5.3* is a possible structure for the Mg^{++} ATP complex.

Table 5.2 which follows lists, together with ATP, the standard free energy of hydrolysis at pH 7·0 for several other biologically important phosphate compounds.

From this short table it is apparent that biological substances can be loosely classified into two categories (according to Jenck's definition) — namely, high and low energy compounds. ATP is seen clearly to occupy what would be termed

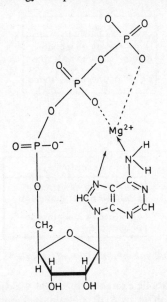

Figure 5.3. Possible structure for the Mg^{++} ATP complex

an intermediary position. Thus regarding their position in the table, those compounds above ATP are known to have a high phosphate group transfer potential, and those below ATP to have a low phosphate group transfer potential.

TABLE 5.2

Substrate	G' at pH 7·0 (cal/mol)
Phosphoenolpyruvate	−12 800
Phosphocreatine	−10 500
Acetylphosphate	−10 100
Mg^{++} ATP complex	− 7 000
Glucose-1-phosphate	− 5 000
Glucose-6-phosphate	− 3 300

ATP, then, is a representative of a common intermediate carrier for biological energy transformation reactions occurring within the cell, for the transfer of phosphate groups from high energy donors to energy acceptors is achieved by the linking ADP–ATP system. It must be appreciated that intracellular reactions

129

proceed via a sequential 'cascade' scheme — that is, the product of one reaction becomes an integral reactant substrate of another, and so on. The 'link' role of the ADP–ATP system is illustrated in *Figure 5.4.*

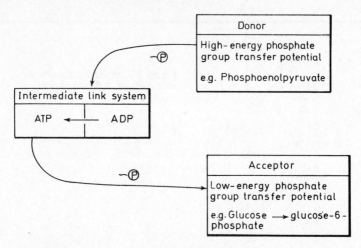

Figure 5.4. Diagram illlustrating 'link' role of the ADP–ATP system

ATP is synthesized in a variety of metabolic exergonic reactions yielding high-energy phosphates which it is able to donate to ADP, thereby recharging it to ATP, and it will be seen in the following sections how energy is conserved as a result of cellular oxidative processes in the form of ATP. Furthermore it will be seen that the ADP–ATP system contributes an important link in co-ordinating the ATP formation and the ATP degradation reactions.

OXIDATION AND REDUCTION

The terms oxidation and reduction may be regarded simply as an expression of the ability of a substance to either lose or gain electrons. The loss of electrons is an important characteristic of oxidation, and such reactions are catalysed by enzymes known as oxidases. Those enzymes which figure predominantly in biological oxidative reactions are the dehydrogenases (*see* Chapter 3, page 92), and these are to be found in the mitochondrial respiratory chain. On the other hand, a substance which acquires electrons is said to have undergone reduction. Hydrogen exhibits a readiness to part with electrons, and for this reason it is a powerful reducing agent, whereas oxygen shows a willingness to gain electrons, and hence it is a strong oxidizing agent.

The reaction: $Zn + Cu^{2+} \rightarrow Zn^{2+} + Cu$ occurs in the Daniell cell so demonstrating a tendency for electrons to leave one solution and enter another, this migration being accompanied by an obvious free energy change. The tendency for a substance to lose or gain electrons will naturally determine the degree to which a system will be oxidized or reduced. This is expressed as the oxidation–reduction potential, E'_o, which is a measurable quantity — the measurement being achieved by using the apparatus shown in *Figure 5.5*. Using such an arrangement, the oxidation–reduction potential of any electrode can be determined relative to the hydrogen electrode ($\frac{1}{2}H_2 \rightarrow H^+ + e$), the potential of which is arbitrarily taken as zero. The two half-cells and their respective electrodes are connected through a potentiometer which will measure the electrical potential difference. The magnitude of this difference is dependent on the oxidation–reduction status of the test solution. The difference of the test compared to the reference is determined by ascertaining the minimum voltage required to prevent an electrical flow. The values have been determined and are known as the electrochemical series.

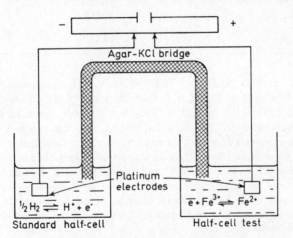

Figure 5.5. Diagrammatic representation of apparatus used for the determination of oxidation–reduction potentials

The stronger the reducing system is the more negative the E'_o is. The difference in oxidation–reduction potential $\Delta E'_o$ between a reducing and an oxidizing agent can be related directly to the free energy change. Thus:

$$\Delta G' = nF\Delta E'_o$$

where $\Delta G'$ is the standard free energy change; n is the number of electrons transferred per molecule; F is the Faraday constant which is equivalent to $-23\,000$ cal/mol; $\Delta E'_o$ the difference in oxidation–reduction potential between an oxidizing and a reducing agent.

Hence electron donors (reducing agents) can be assembled into a thermo-dynamic series by virtue of their decreasing oxidation—reduction potential, which clearly illustrates the tendency for electrons to flow from the most negative to the most positive in the scale (*see* Table 5.3).

TABLE 5.3

Redox system	E'_0 (volts)	$\Delta G'$ (cal/mol)	Thermodynamic direction
NAD	−0·32	−12 400	
Flavin	−0·20	− 4 100	
Cytochrome b	+ 0·12		
Cytochrome c	+ 0·22	−10 000	
Cytochrome a	+ 0·29	− 1 380	
$H_2O \rightarrow H_2 + O_2$	+ 0·815	−24 400	
Total		−52 180	

Thus the reaction $NADH_2 + \frac{1}{2}O_2 \rightarrow NAD + H_2O$ proceeds with a large decrease in free energy — that is, it is strongly exergonic. In fact ΔG is in the region of −52 000 cal/mol. But $NADH_2$ is not rapidly oxidized by molecular oxygen as suggested by the formula, for in fact the reaction requires the participation of several enzymes, the sequence being summarized in Table 5.3.

A large proportion of the metabolic activity of an organism is thus devoted to the generation of ATP from ADP and phosphate. The means by which an organism attains its energy is characteristic of its physiology —and the main distinction to be made here is the recognition of two major forms of metabolic activity — that is, aerobic and anaerobic respiration, of which the latter will now be considered.

ANAEROBIC RESPIRATION

Within the biosphere the vast majority of resident organisms achieve an efficient release of energy in the presence of oxygen, and in such cells molecular oxygen constitutes the ultimate electron acceptor. On the other hand there are a few organisms which are strictly anaerobic in which an inorganic compound other than oxygen is the ultimate electron acceptor. These anaerobic cells are micro-organisms — for example, the Clostridia genera. In fact the bacteria are seen to form three groups with regard to their ability to respire in either the presence or absence of oxygen: (1) obligate aerobes which are dependent on oxygen; (2) obligate anaerobes, which are cells capable of growth only in the absence of oxygen, though a few may be micro-aerophilic — that is, able to tolerate the presence of oxygen at a reduced tension; (3) facultative anaerobes — cells normally requiring oxygen, but which exhibit the ability to exist under anaerobic conditions by shifting to an alternative respiratory mechanism as utilized by the strict anaerobes. Such cells, in result, tend to have a greater metabolic activity. In most cases of facultative cells the means of obtaining energy is from the degradation of glucose, by a process known as anaerobic glycolysis. The breakdown of a single glucose molecule in a facultative anaerobic cell yields two lactic acid molecules; under aerobic conditions the end-product would have been carbon dioxide and water. Further to this, facultative cells will yield quite different end-products of fermentation, which will be determined by the type of organism involved. Hence the apparently wide variety of fermentation pathways are all fundamentally based on the Embden—Meyerhof pathway — which is illustrated diagrammatically in *Figure 5.6* — up to the formation of pyruvic acid; at this stage, under anaerobic conditions, it can be subsequently converted to lactic acid or ethanol. In the presence of oxygen (aerobic respiration) it will enter the aerobic or Krebs cycle (*see* page 146), to complete the oxidation of the molecule to water and carbon dioxide.

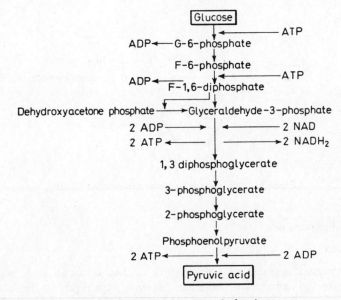

Figure 5.6. The Embden—Meyerhof pathway

In the glycolytic formation of pyruvic acid the generation of ATP is dependent on the oxidation of triose phosphate at the expense of the reduction of nicotinamide adenine dinucleotide (NAD) (*see* Photosynthesis, page 147).

Following this, the reduced NADH donates electrons to pyruvic acid thereby regenerating NAD. If this re-oxidation were not carried out rapidly and efficiently, fermentation within the cell would cease, since the total NAD content of the cell is remarkably small. In mammalian tissue nicotinamide is found as a constituent of the nucleotide NAD, and is a vitamin of the B class. It is an essential dietary component the absence of which will precipitate the clinical syndrome, pellagra. The structure of NAD is illustrated in *Figure 5.7.*

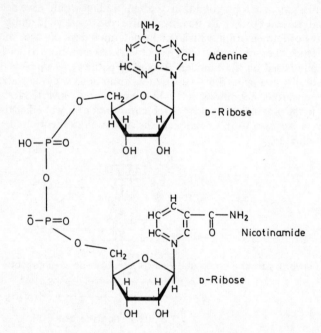

Figure 5.7. Nicotinamide adenine dinucleotide (NAD)

Types of Fermentation

Lactic Acid

The simplest example of fermentation is that of lactic acid, and it is characteristic of mammalian tissue and bacteria. With regard to the latter this form of fermentation is particularly characteristic of the lactic acid bacteria — for example, the Streptoccoci and the Lactobacilli. In this type of fermentation, hexoses are converted into two trioses, the result being attained by the breakdown of two molecules of ATP. The oxidation of triose phosphate is coupled with the reduction of pyruvic acid through the participation of NAD. Though four ATP molecules are subsequently produced as a result of this fermentation pathway, the net gain is 2 ATP molecules per hexose. Lactic acid fermentation is compared with alcoholic fermentation in *Figure 5.8.*

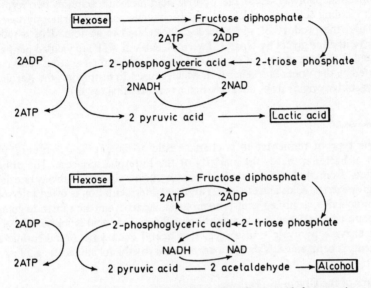

Figure 5.8. Comparative metabolic pathways: lactic acid fermentation compared with alcohol fermentation

Alcoholic

This pathway is characteristic of the higher plants, yeasts and certain fungi; it is uncommon in the bacteria, though an exception is the Zymomonas. The major difference between alcoholic and lactic acid fermentation is seen to be that the pyruvic acid in the latter case is de-carboxylated to carbon dioxide and acetaldehyde this then being reduced to alcohol in an NAD linked reaction (*Figure 5.8*).

Propionic Acid

This particular type of fermentation is exhibited by the propionibacteria which are micro-aerophilic and anaerobic organisms found inhabiting the intestinal tract of cattle. Here the pyruvic acid is carboxylated to oxaloacetic acid which is then reduced to succinic acid, this being finally de-carboxylated to yield propionic acid. As a result of this pathway only one ATP molecule is formed for every 9 carbons fermented.

Butylene Glycol

The Aerobacter and related organisms are responsible for this type of fermentation pathway. Here the pyruvic acid molecule acquires two hydrogen atoms yielding active acetaldehyde which condenses with a further pyruvic acid molecule, the product of which is de-carboxylated to acetoin. This is reduced to 2, 3-butylene glycol by NADH. Two molecules of ATP are yielded per hexose donated. Butylene glycol on exposure to air is oxidized to acetoin. This can be detected by the Voges-Proskauer test, which is used to distinguish the Aerobacter (of vegetative origin) from the Escherichia (of intestinal origin).

Formic Acid

This type of fermentation is characteristic of the enteric-dysentery-typhoid group of bacteria; that is the majority of the Enterobacteriaceae. The principal products from the fermentation of hexose sugars via this pathway are lactic, formic and acetic acids and ethanol. This type of fermentation is often referred to, in consequence, as mixed acid fermentation. Apart from the substrate going via the lactic acid pathway, a major portion of the pyruvic acid is split by the action of an enzyme pyruvate transacetylase to acetyl coenzyme A and formic acid. The acetyl coenzyme A is then in part reduced to ethanol and acetic acid.

Features of the Glycolytic Pathway

The glycolytic pathway commences with Stage 1 and the phosphorylation of D-glucose to D-glucose-6-phosphate by the action of the enzyme hexokinase (ATP: D-hexose-6-phosphotransferase); this is then converted to D-fructose-6-phosphate in Stage 2, by the enzyme phosphoglucomutase. Stage 3 represents that point at which another molecule of ATP (ATP: D-fructose-6-P-1-phosphotransferase) is consumed in the conversion of fructose-6-phosphate to fructose-1, 6-diphosphate. Next follows the conversion of fructose-1, 6-diphosphate into two molecules of triose phosphate and the interconversion of aldo and ketotriose phosphates — that is, Stages 4 and 5 respectively, with 3-phosphoglyceraldehyde and dihydroxyacetone phosphate in result. Stage 4 is catalysed by the enzyme aldolase, whereas stage 5 is an interconvertible and reversible reaction in which the participating enzyme is triose phosphate isomerase. The earlier input of ATP is seen to be recovered in the following chemical sequence — that is, after the phosphorylation of ADP at the expense of dehydrogenation of glyceraldehyde phosphate, which donates electrons to the oxidized form of NAD thereby reducing it. The enzyme catalysing Stage 6 is glyceraldehyde phosphate dehydrogenase. At this point 2ADP molecules are regenerated to ATP (Stage 7). This occurs when 1, 3-diphosphoglycerate is converted to 3-phosphoglycerate. The Stage 7 reaction is catalysed by diphosphoglycerate kinase (ATP: 3-phospho-D-glycerate-1-phosphotransferase). Stage 8 is a reversible, interconverting

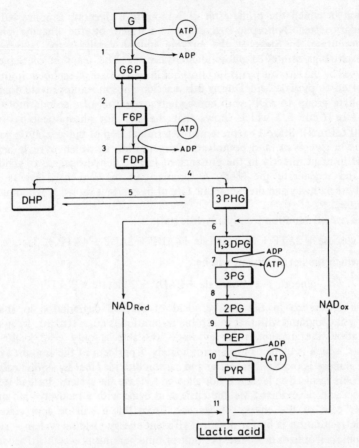

Enzymes

(1) Hexokinase; (2) phosphoglucomutase; (3) phosphofructase; (4) aldo-lase; (5) triose phosphate isomerase; (6) glyceraldehyde phosphate dehydrogenase; (7) diphosphoglycerate kinase; (8) phosphoglyceromutase; (9) enolase; (10) pyruvate phosphokinase; (11) lactate dehydrogenase

Intermediates

G: D-glucose; G6P: glucose-6-phosphate; F6P: fructose-6-phosphate; FDP: fructose-1, 6-diphosphate; DHP: dihydroxyacetone phosphate; 3PHG: 3-phosphoglyceraldehyde; 1, 3DPG: 1, 3-diphosphoglycerate; 3PG: 3-phosphoglycerate; 2PG: 2-phosphoglycerate; PEP: phosphoenolpyruvate; PYR; pyruvic acid

Figure 5.9. Diagrammatic summary of the major events in the glycolytic pathway

137

reaction in which the prime ester of 3-phospho-D-glycerate is converted to the secondary ester 2-phospho-D-glycerate, catalysed by the enzyme phosphoglyceromutase. In Stage 9, the enzyme enolase catalyses the conversion of 2-phospho-D-glycerate to phosphoenolpyruvate. The reaction of Stage 10 is catalysed by the enzyme pyruvate phosphokinase — where phosphoenolpyruvate is converted to pyruvic acid. During this reaction phosphoenolpyruvate donates its phosphate group to ADP, with consequent generation of a second molecule of ATP (*see* Table 5.2 which shows that the $\Delta G'$ for phosphoenolpyruvate is 13 000 cal/mol). Stage 11 represents the terminal step of the glycolytic pathway in muscle glycolysis, and homolactic fermentation in which pyruvic acid and NADH interact directly in the presence of lactate dehydrogenase to yield lactic acid and regenerate the NAD. A summary of the sequential events of the glycolytic pathway and the metabolic fate of pyruvate is shown diagrammatically in *Figure 5.9.*

The overall equation for glycolysis is thus:

glucose + 2ATP + 2 phosphate +4ADP → 2ADP + 4ATP +2 lactate

from which the net statement will be:

glucose + 2 phosphate + 2ADP → 2 lactate + 2 ATP

The difference in the energy yield of glucose degradation in anaerobic glycolysis compared with that in aerobic respiration is quite striking. In anaerobic respiration the maximum energy made available is only −52 kcal/mol of glucose, which is in fact only approximately 7 per cent of the amount available when glucose is oxidized to water and carbon dioxide (that is, −686 kcals/mol). In aerobic respiration lactate is not allowed to leave the system: instead it, or the pyruvic acid, is oxidized via the citric acid cycle with a recovery rate of some 90 per cent of the energy of glucose. From this it will be appreciated that aerobic respiration is by far the most efficient energy-yielding system — since an anaerobic cell utilizes more fuel per unit of time per unit of weight to accomplish an amount of cellular work comparable to that manifested by an aerobic cell. The alternative metabolic routes taken by certain micro-organisms at the terminal stages of the glycolytic pathway are illustrated in *Figure 5.10.*

Alternative Pathways of Hexose Degradation

Not all facultative anaerobic cells proceed via the Embden-Meyerhof pathway of glycolysis. For example *Leuconostoc mesenteroides,* the heterolactic fermenters, yield carbon dioxide, lactic acid and ethanol via a pathway called the pentose-P-phosphoketolase pathway, in which carbon dioxide is formed by the decarboxylation of phosphogluconate; and ethanol as the result of acetyl phosphate reduction, catalysed by the enzyme phosphoketolase. Similarly the Pseudomonads have established an alternative route called the Entner-Doudoroff pathway, in which the oxidation of glucose-6-phosphate to 6-P-gluconate is

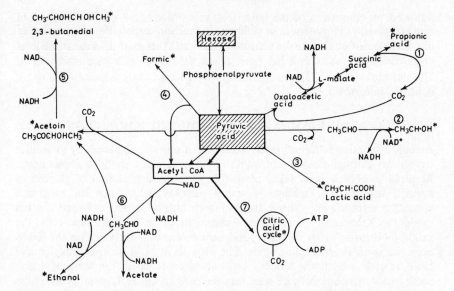

Figure 5.10. Alternative fates of pyruvic acid – the terminal stages of the glycolytic pathway. (1) Propionibacteria and Veillonella; (2) yeasts; (3) Lactobacillaceae; (4) Enterobacteriaceae (e.g. Escherichia coli); (5) Aerobacter; (6) some Enterobacteriaceae and Clostridium; (7) general direction of mammalian pathway (see Figure 5.11)

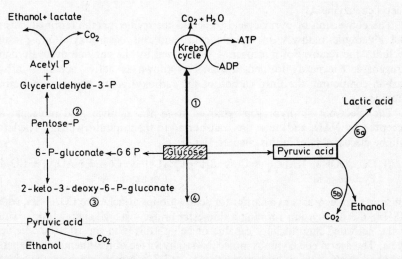

Figure 5.11. Alternative metabolic routes: (1) mammalian; (2) pentose-P-phosphoketolase pathway, taken by Lactobacilli and Leuconostoc mesenteroides; (3) Entner–Doudoroff pathway – pseudomonas; (4) the pentose-P pathway; (5) Embden-Meyerhof pathway: (a) mammalian muscle, and the Lactobacillaceae; (b) anaerobic yeasts

followed by conversion of the latter to 2-keto-3-deoxy-6-P-gluconate, which in turn is converted to pyruvic acid yielding ethanol and carbon dioxide. Further to this, some organisms utilize the pentose-P-pathway. This route diverges at the level of glucose-6-phosphate and has been called the hexose monophosphate shunt. The patterns of the alternative pathways which are briefly mentioned here are indicated in *Figure 5.11*.

AEROBIC RESPIRATION AND THE MITOCHONDRION

Aerobic respiration is the enzymatic oxidation of organic substances by molecular oxygen to carbon dioxide and water; this takes place in the mitochondria. Respiration of this type takes place following the degradation of glucose to pyruvic acid via the Embden–Meyerhof pathway, whereupon the substance enters the aerobic cycle. This is the metabolic hub of aerobic cells and is often called the Krebs citric acid cycle. Fundamentally in heterotrophic cells, the major foodstuffs – fats, proteins, and carbohydrates – converge on the Krebs cycle via specific metabolic routes. The citric acid cycle, or tricarboxylic acid cycle as it is sometimes termed, can only accept fuel in the form of acetyl coenzyme A. Hence it will be seen that the carbohydrates, fats and proteins are initially degradated to pyruvic acid which is then converted to acetyl coenzyme A. The fatty acids and amino acids are able to enter the citric acid cycle via alternative routes (*see* Homoeostasis, page   ). To simplify matters, the cycle will be dealt with commencing with the stage at which pyruvic acid is converted to acetyl coenzyme A.

The conversion of pyruvic acid to acteyl coenzyme A is the terminal stage of the glycolytic pathway in aerobic cells, and this decarboxylation – which includes the removal of hydrogen – is achieved by the enzyme pyruvate dehydrogenase. The newly formed substance is known as 'active' acetate – a two-carbon compound, the third carbon of the carboxyl group being lost as carbon dioxide.

The electrons of hydrogen removed from the pyruvic acid molecule are accepted by NAD, and these are transported to the respiratory chain. The latter will be discussed later in this chapter. Hence:

$$CH_3CO.COOH + NAD_{ox} + CoA{-}SH \rightarrow CH_3CO{-}S{-}CoA + NAD_{red} + CO_2 \uparrow$$
$$\text{Pyruvic acid} \qquad\qquad\qquad\qquad \text{Coenzyme A}$$

This coenzyme A acts as a carrier for acetyl groups attached to COA–SH, where –SH is the thiol group forming a thio-ester bridge – a high energy link – which is the activated intermediary capable of being utilized in the tricarboxylic acid cycle. The acetyl coenzyme A molecule has, by virtue of its inherent high-energy linkage, a high $-\Delta G'$ of hydrolysis; that is, -7.7 kcal/mol, which can be equated with that of the ATP system.

Thus with the formation of acetyl coenzyme A, the molecule then donates its acetyl group to the four-carbon dicarboxylic acid, oxaloacetic acid to form the condensate citric acid:

$$CH_3CO-S-CoA + \text{oxaloacetic acid} + H_2O \rightarrow \text{citric acid} + CoASH + H^+$$

With the formation of citric acid, coenzyme A is regenerated, thereby forming a dynamic oxidation cycle, such that the citric acid cycle is constantly fed with active acetate molecules. This first reaction is catalysed by the enzyme known as the citrate-condensing enzyme or citrogenase. Next the enzyme aconitase catalyses the conversion of citric acid into two tricarboxylic acids — cis-aconitic and isocitric acids. There exists between the citric, cis-aconitic and isocitric acids a reversible hydration equilibrium. This conversion of citric acid to cis-aconitic acid is followed by the removal of water molecules, and a subsequent uptake of water in the conversion reaction of cis-aconitic to isocitric acid which follows. Isocitric acid is then converted to the five-carbon acid, α-ketoglutaric acid, the reaction being catalysed by the enzyme isocitrate dehydrogenase — thus:

$$\text{isocitric acid} + NAD \rightarrow \alpha\text{-ketoglutaric acid} + NADH + CO_2 \uparrow$$

As the above equation indicates, two atoms of hydrogen are removed and transported to the respiratory chain by NAD and CO_2 is evolved. A further decarboxylation–dehydrogenation step then follows:

(1) α-ketoglutaric acid$^-$ + CoASH + NAD $\rightarrow$ succinyl-S-CoA$^-$ + NADH + CO_2 ↑
(2) succinyl-S-CoA + GDP + P $\leftrightharpoons$ succinic acid + GTP + CoA-SH

in which the five-carbon compound α-ketoglutaric acid is oxidized to the four-carbon compound succinic acid and carbon dioxide — catalysed by the enzyme α-ketoglutaric dehydrogenase, the hydrogen being accepted by NAD, which is thereby reduced. In fact, as implied above, during this transition a product succinyl CoA is formed the preponderant fate of which is the formation of succinic acid. The reaction, which is a highly exergonic one, is catalysed by the enzyme succinic thiokinase. The next step is the dehydrogenation of succinic acid to fumaric acid by the enzyme succinic dehydrogenase:

$$\text{succinic acid} + FAD \leftrightharpoons \text{fumaric acid} + FADH_2$$

The fumaric acid is then converted to malic acid (a reversible hydration reaction catalysed by the enzyme fumarate hydratase):

$$\text{fumaric acid} + H_2O \rightleftharpoons \text{malic acid}$$

The formed malic acid is then dehydrogenated to oxaloacetic acid by the enzyme malate dehydrogenase:

$$\text{malic acid} + NAD^+ \leftrightharpoons \text{oxaloacetic acid} + NADH + H^+$$

NAD being the electron acceptor.

Thus the cycle is completed with the regeneration of oxaloacetic acid, the

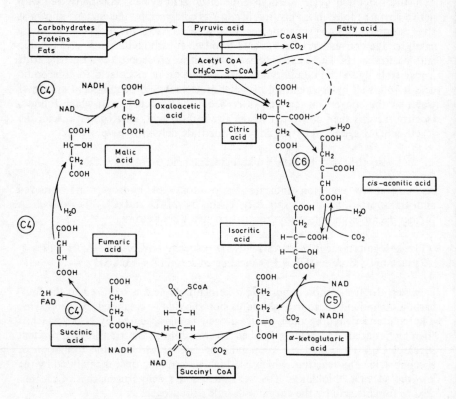

Figure 5.12. The citric acid cycle

initial substrate material. A summary of the major events of the citric acid cycle briefly surveyed here is given in *Figure 5.12*.

From the text above and *Figure 5.12* it is evident that although compounds of six, five and four carbon atoms participate in the cycle, all can be regenerated so long as activated acetate, together with a suitable electron acceptor, is available to the system. Though the reactions of the citric acid cycle serve as a source of intermediates for the biosynthesis of amino acids (*see* Homoeostatic Function), the cycle also serves as a catabolic and energy-generating device — namely, in the formation of ATP, which is primarily achieved by the intimate association of the cycle with the respiratory chain, and the process known as oxidative phosphorylation.

Clearly, then, in one revolution of the cycle, there are four dehydrogenation steps. Three of these involve NAD as the electron acceptor and the other FAD, (flavin adenine dinucleotide). These electrons are then donated to a series of enzymes forming the respiratory chain which forms the final common metabolic pathway, via which electron derivatives of biological fuels flow to oxygen.

The general concept that oxidations were mediated by electron carrier groups in a chain of ascending redox potential was formulated from the work of Keilen in 1929 and of Warburg in 1932. The sequence of these respiratory components — the pyridine-attached dehydrogenases, the flavoproteins and the cytochromes — was deduced by Chance (1956). It was found that these components were arranged in specific sequence of spatially related assemblies on the shelves and inner membranes of the mitochondrial wall. The cytochromes are haem groups consisting of porphyrin and iron. The iron atoms of each cytochrome molecule can exist either in the Fe^{++} (reduced) or the Fe^{+++} (oxidized) form. Most of the components of the respiratory chain are found in the form of lipoprotein complexes that can be derived from the inner membrane of the mitochondrion (Green, 1962), and four such lipoprotein complexes have been recognized and are designated I–IV. The lipoprotein complex I contains flavin and a non-haem iron. Complex II contains flavin also — but unlike I it is not acid extractable — together with a non-haem iron. The lipoprotein complex III contains cytochrome b and c, and complex IV possesses cytochrome a,a_3 and copper ions. Thus with the exception of ubiquinone (coenzyme Q or CoQ) and cytochrome c, all the others are complexed with protein, and firmly bound to the inner mitochondrial membrane. It has been shown that the lipids are essential components of the electron transport system (Green and Fleischer, 1960). The spatial arrangement of the respiratory chain components is shown diagrammatically in *Figure 5.13*.

Fundamentally electrons derived from the citric acid cycle and the fatty acid β-oxidation cycle enter the cytochrome chain via either NAD or FAD. It is considered (Green, 1962) that complex I accepts electrons from NADH and complex II removes them from succinic acid (*Figure 5.13*). Both of these can donate their electrons to cytochrome b via the common intermediary ubiquinone (CoQ). Some workers, however, are inclined to believe CoQ is not an obligatory component of the electron chain; whether this is or is not true Complex III receives electrons from complexes I and II, and then transfers them to cytochrome c. The resultant reduction then proceeds to complex IV — water being the net product. Hence the overall reaction of the oxidation of NADH by molecular oxygen is:

$$NADH_2 + \tfrac{1}{2}O_2 \rightarrow NAD + H_2O$$

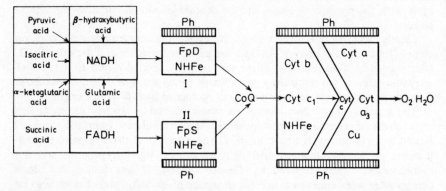

NADH: reduced nicotinamide-adenine dinucleotide; FADH: reduced flavin-adenine dinucleotide; FPD: NADH dehydrogenase – an acid-extractable flavin; Fps: succinic dehydrogenase – an acid non-extractable flavin; NHFe: non-haem iron; CoQ: Coenzyme Q – ubiquinone, with a long polyisoprenoid side chain; Cyt a, b, c_1 and c, a_3: respective cytochromes; Cu: copper; Ph: phospholipids

Figure 5.13. Diagrammatic representation of the respiratory chain

which proceeds with a large decrease in ΔG. The thermodynamic relationships in the respiratory chain are shown in Table 5.4.

TABLE 5.4

	E'_o (mV)	$\Delta E'_o$ (mV)	ΔG° (cal)
NAD	−320	+257	−118500
Flavoprotein	− 63	+ 23	− 1060
Cytochrome b	− 40	+290	− 13370
Cytochrome a	+250	+ 40	− 1840
Oxygen	+815	+525	− 24210
		Total ΔG°	− 52320

The release of energy during the oxidation of the electron pairs moving down the respiratory chain is coupled with the synthesis of ATP. This transfer of electrons along the respiratory chain and the coupling of the transfer to the phosphorylation of ATP, is called oxidative phosphorylation. In fact during this process 3 molecules of ATP are formed as a result of oxidation of the 4 substrates of the Krebs cycle — in which a pair of electrons from each are supplied to the respiratory chain, and ultimately reduce one atom of oxygen. Within the chain there are three areas in which a large free energy change occurs: (1) at NAD to flavoprotein; (2) cytochrome b to cytochrome c; and (3) cytochrome a to oxygen (Table 5.4), and it is at these sites that high energy intermediates are formed, and subsequently donated to ATP. The precise mechanisms of these coupling reactions are not known, but Figure 5.14 illustrates a proposed model. Thus, as a result of oxidation phosphorylation, approximately twenty times more energy is released than that yielded by the degradation of glucose to pyruvic acid, and thermodynamically the efficiency of the process is approximately 50 per cent. The remaining 50 per cent is lost to the system as heat.

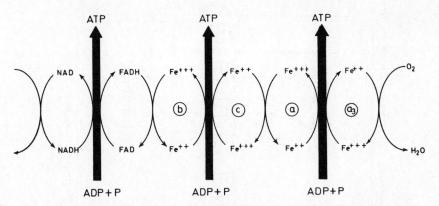

Figure 5.14. Schematic plan of coupling reaction of the respiratory chain

The control of the metabolic activity of the organism is discussed later in Chapter 8 which deals with homoeostatic function. In general, however, it may be said that metabolism is largely self-controlling. The utilization of ATP in a particular cellular activity will precipitate an increase in ADP and inorganic phosphate, and by mass action the resultant concentration will motivate their conversion to ATP, thus promoting a continuous cycle of degradation and synthesis, requiring in consequence an equally continuous supply of raw materials.

Finally, as implied earlier, the precise mechanism of oxidation phosphorylation has yet to be elucidated; this should eventually emerge from the present study of various mitochondrial structures such as those isolated from lower organisms, yeasts and mammalian tissue and by the use of artificial membrane models. A diagram linking citric acid cycle and respiratory chain is shown in *Figure 5.15.*

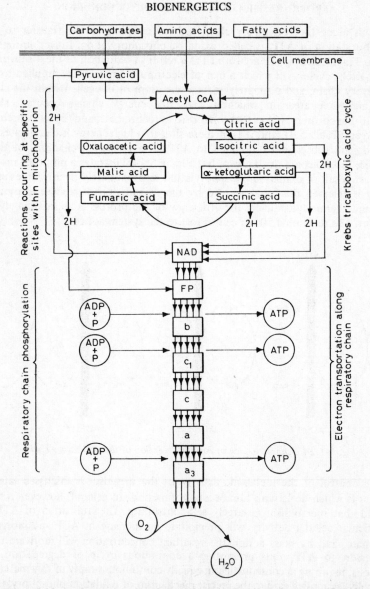

Figure 5.15. Oxidative phosphorylation: diagram depicting the energy flow from the Krebs cycle along the respiratory chain within the mitochondrial cristae, and its conservation in the form of ATP. The respiratory chain is the common final pathway, via which electron derivatives of biological fuels flow to oxygen (after Spencer, 1970; reproduced by courtesy of Butterworths; London)

PHOTOSYNTHESIS

The *status quo* of the biosphere is dependent on the ability of plants (this includes the higher plants, ferns, mosses, diatoms, the green, blue-green, red and brown algae), and also some micro-organisms such as Euglena, the Thiorhodaceae, Athiorhodaceae, and the Chlorobacteriaceae to: (1) convert solar energy into the chemical energy which is required to drive endergonic cellular reactions synthesizing carbohydrates from carbon dioxide and water; (2) restore environmental balance by the consumption of carbon dioxide and the generation of oxygen; (3) provide essential foodstuffs for the heterotrophic members of the biosphere; (4) academically satisfy the second law of thermodynamics.

The overall stoichiometry of photosynthesis may be represented thus:

$$6\,CO_2 + 6H_2O + h\nu \rightarrow C_6H_{12}O_6 + 6O_2$$

which demands that for each volume of carbon dioxide absorbed an equal volume of oxygen should be liberated, hence the photosynthetic quotient (Q_p) should be unity:

$$Q_p = \Delta O_2 / -\Delta CO_2$$

The way in which carbon dioxide is incorporated into a light-energized system, so as to produce high-energy-yielding compounds has been virtually solved in terms of an overall conception (Calvin, 1962), though there remain areas of this process which are still uncertain — particularly when considering the physicochemical mechanisms governing the excitation of the chlorophyll molecule by light within the lamellate grana.

Speculation and experimentation in this field extends back to Priestley (1771), who discovered that during daylight plants expired oxygen. Towards the end of the nineteenth century Winogradsky demonstrated that certain bacteria (later identified as the Nitrosomonas) were able to assimilate carbon dioxide in the dark without the aid of chlorophyll. In fact it was not until the mid-twentieth century that workers finally appreciated fully the existence of 'light' and 'dark' reactions. For though the term photosynthesis implies the participation of light, and is used generally to describe the total process by which carbon dioxide and water are converted to glucose and oxygen at the expense of solar energy — light is only directly involved during the initial so-called 'light' reaction. Once light has been trapped, and transformed into chemical energy, the subsequent conversion reactions proceed without the participation of light — hence they are called the dark reactions.

The true nature of light (solar radiation) is not fully understood. In order to comprehend and account for the phenomena associated with light energy, two theories have to be considered in relation to one another, one being the wave theory and the other the quantum theory. Fundamentally the wave theory regards light as electromagnetic radiation, and the quantum theory visualizes light as consisting of particles known as photons or quanta.

Electronic Energy

The energy of a molecule may be divided into different levels. The energy of electrons in the atom described by the quantum numbers – in other words the energy associated with the transition or promotion of electrons from one energy state to another is termed electronic. As postulated by Bohr, the total electron component of an atom is arranged in a number of different energy levels, with the electron following a particular trajectory about the nucleus – which possesses a particular quantity of potential energy. The effect of movement of electrons from one level to another is manifested by either absorption or loss of energy according to the direction of electron movement. The result of electron promotion is illustrated in *Figure 5.16*. From this illustration it will be seen that the shell or orbital closest to the nucleus is associated with the lowest energy state.

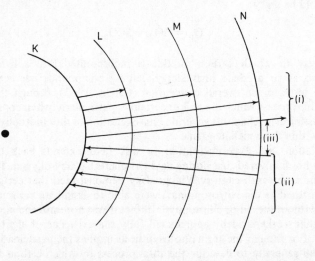

Figure 5.16. Diagram showing electron promotion – associated with a gain or loss of energy by an atom: (i) energy absorbed; (ii) energy released; (iii) indicating direction of electron promotion

Internal Energy

This is a term embracing vibrations of the constituent atom, and rotational energy which consists of atomic rotations about an axis. Thus the compound term internal energy is associated with the phenomenon of emission and absorption spectra of infra-red rays. In contrast, changes in electronic energy are followed by the emission of ultra-violet light or visible light.

Prior to the development of the quantum theory it had been assumed that energy emitted from a body was in a continuous stream, and that the frequency of the emission was determined by the nature of the radiation — for example, heat or light. Planck had visualized a discontinuous emission of energy, which he called quanta — possessing a magnitude of hv, where v is the frequency of vibration and h is a constant value for all frequencies (6.5×10^{-27} ergs/sec). Thus energy was seen to be quantized, such energy being expressed by the formula nhv, where n is an integer, and can be represented diagrammatically as follows:

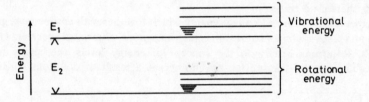

Thus when light is absorbed by a molecule, energy is in effect absorbed, and this energy must be manifested in an energy transition within the molecule:

$$E_2 \rightarrow E_1, \text{ then } E_1 - E_2 = hv$$

It was Einstein who proposed that photons travelling through space were in fact units of energy or light quanta of magnitude hv. Thus from the above, the energy of light of a particular wavelength is given by the equation:

$$E = nhv$$

Hence the shorter the wavelength, the greater the frequency of the emission and energy in one photon.

Note that:

$$c = v\lambda$$

where c is the velocity of light (3×10^{10} cm/sec) and λ is the wavelength.

Thus we can write:

$$E = hv = h.c/\lambda$$

where E is the energy of light quanta, v is the frequency of vibration, h is the Planck constant, c is the velocity of light and λ is the wavelength. Since in photosynthesis one quanta of radiation activates one molecule of the reacting substances, then that energy required to activate one gramme-molecule of the substance is Nhv, where N is the Avogadro number. The Avogadro number, it will be remembered, is the number of molecules in a gramme molecular weight of a substance at $0°C$ and 760 mm pressure (that is, 6.023×10^{23}). This amount of light energy is called an einstein. Thus one einstein is Nhc/λ, where N is 6.023×10^{23}, h is 6.624×10^{-27}, and c is 2.997×10^{10} cm/sec — the energy being expressed in ergs; if divided by 4.184×10^{14}, the resultant figure is in calories. Further to this, according to Einstein's photochemical equation:

$$\tfrac{1}{2}\,mv^2 = hv - \phi$$

where $\tfrac{1}{2}mv^2$ is the kinetic energy of the photochemical mass (m), with a velocity (v), and ϕ is the amount of work done as a result of emission. Thus if $hv=\phi$, there is no emission.

Light is thus a form of radiant energy which together with cosmic rays, gamma rays and x-rays, etc constitutes the electromagnetic vibrational spectrum (*Figure 5.17*). Returning briefly to the concept of energy levels, the energy of one molecule may be transmitted to another as a result of collision. Earlier, on

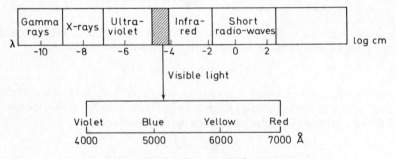

Figure 5.17. The electromagnetic spectrum

page 124 it was seen that some reactions will proceed spontaneously, whereas others require that a quantity of energy be put into the system prior to the reaction proceeding to completion, this being termed activation energy. This acquisition of activation energy may result from thermal collision, motivating an acceleration in the reaction rate. In photochemical reactions this activation energy is supplied in the form of visible or ultra-violet light, and a molecule will react each time it has absorbed a quantum of light (on an all-or-none basis, since a part of a quantum cannot be utilized) — providing this energy value is equivalent to or greater than the activation energy.

Referring again now to *Figure 5.17,* it will be seen that some rays have short and others long wavelengths. Those with long wavelengths have correspondingly low frequencies and low energy quanta. Conversely, the short wavelengths have a high frequency and a reciprocating high energy content. The far ultra-violet portion of the spectrum is the region of characteristic absorption by simple molecules, and the near infra-red is the region of vibrational and rotational (but not electronic) transitions.

When light is absorbed, the molecule responsible for the absorption is excited to a higher energy state in which it remains for approximately 10^{-8} seconds. The excited molecule can lose its energy, thus returning to the ground state in a variety of ways: (1) by dissociation of the molecule; (2) as heat; (3) as light of a different wavelength — that is, fluorescence; (4) as light of a different wavelength after a time lag — phosphorescence (*Figure 5.18*).

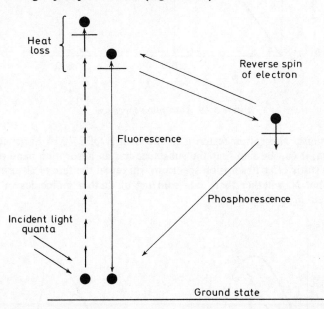

Figure 5.18. Excitation of molecules to fluorescence or phosphorescence

The amount of light as a function of wavelength is known as the absorption spectrum; by virtue of its dependence on chemical structure, different spectra are obtained with each individual structure. Measurement of the absorption spectrum can be made utilizing Beer's law which states:

$$D = \log_{10}(I_0/I_t) = E.c.l$$

where D is the optical density; I_0 is the intensity of the radiant light; I_t is the intensity of the transmitted light; E is the molar extinction, which is a function of wavelength; c is the concentration of the test solution in mols per litre and l is the length of the cell.

From this it can be seen that for a fixed concentration and cell size, E will determine the extent of absorption. Hence by plotting the extinction coefficient against wavelength, an absorption curve of spectra is obtained. In *Figure 5.19,*

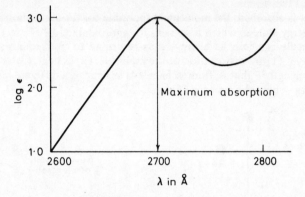

Figure 5.19. Absorption spectrum

the most intense absorption region is in the vicinity of 2 700Å λ; hence in the curve shown, it can be said that the substance has an absorption band of 2 700 Å. Thus the study of an absorption spectrum will reveal the fate of absorbed light energy — that is, whether there is a splitting of diatom molocules or if they simply remain in an excited state.

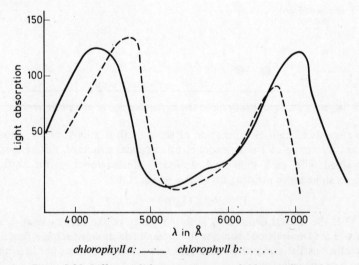

chlorophyll a: ———   chlorophyll b: ......

Figure 5.20. Differential absorption spectra of chlorophyll a and b

Chlorophyll and the Photochemical Reactions in Photosynthesis

As mentioned in Chapter 3, page 91, two kinds of chlorophyll, namely a and b are found in the higher plants; these differ marginally both structurally and in absorption spectra. The structure of chlorophyll a which consists of a tetrapyrrole ring centred about a magnesium nucleus, and a long phytyl side chain is shown in *Figure 4.4.* By contrast chlorophyll b, contains the —CHO (formyl) group instead of a methyl group in one of the pyrrole rings. The differential absorption spectra of chlorophylls a and b are shown in *Figure 5.20;* this reveals clearly for both, that maximum absorption occurs in the red and blue bands — and it is for this reason that leaves look green.

Isolated chlorophyll molecules become excited *in vitro* when exposed to light — returning to the ground state by the loss of light energy as a result of heat dissipation and fluorescence (*Figure 5.18*). In 1939, Hill demonstrated that an illuminated suspension of chloroplasts in water, to which hydrogen acceptors were added evolved oxygen without synthesizing carbohydrates. This Hill reaction consists of the photolysis of water:

$$2H_2O \rightarrow 4H^+ + O_2 \uparrow + 4e^+$$
$$\underline{4\ Fe^{+++} + 4e^- \rightarrow 4Fe^+}$$
$$4\ Fe^{+++} + 2H_2O \rightarrow 4H^+ + O_2 \uparrow + 4Fe^{++}$$

and essentially it is the presence of the electron acceptor (a ferric salt) which permits the reaction to proceed. The chlorophyll molecule, *in vivo*, is excited by light, giving rise to high-energy electrons which in consequence become involved in a cyclic pathway permitting the conservation of some of this energy in a chemical form rather than its loss in such forms as fluorescence. At present this photochemical reaction is not fully understood, but it has been explained as being akin to the coupling principle mechanism occurring in the respiratory chain in the mitochondrion. There exists, then, a cyclic electron transport pathway enabling the phosphorylation of ADP to ATP to take place by a process called cyclic photophosphorylation, and also the reduction of NADP to NADPH to be effected. *Figure 5.21* shows for each pair of electrons donated to this pathway, 2 ATP molecules are formed from ADP and phosphate. But this is only half the story, for this initial photochemical reaction also brings about the hydrolysis and ionization of water in which NADP is reduced to NADPH. NADP is nicotinamide adenine dinucleotide phosphate, which is virtually identical to NAD except that it has an additional phosphate group attached to the D-ribose ring at the 2-hydroxyl group.

It is believed that when chlorophyll a becomes excited by absorbing the far red light quanta, an electron with a high reducing potential converts NADP to NADPH; the electron from chlorophyll b (excited) restores chlorophyll a to its ground state. At the same time chlorophyll b donates electrons to the cyclic

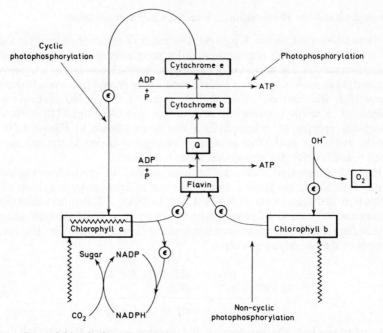

Figure 5.21. The first stage of photosynthesis: the photochemical reaction

photophosphorylation chain. Thus as electrons flow from b to a chlorophyll ADP is phosphorylated to ATP, by a process known as non-cyclic photophosphorylation. Hence in cyclic photophosphorylation the pathway commences and terminates with chlorophyll a, whereas in the non-cyclic pathway it goes from b to a. The light phase of photosynthesis – the photolysis of water resulting in the formation of ATP and the reduction of NADP – is summarized in *Figure 5.21*.

The Dark Thermochemical Reaction in Photosynthesis: The Calvin–Bassham Cycle

Glucogenesis commences with the combination of atmospheric carbon dioxide with ribulose–1, 5-diphosphate to form two molecules of 3-phosphoglycerate, which undergoes immediate enzymatic reduction by triose phosphate dehydrogenase, at the expense of ATP and NADPH, to yield two molecules of 3-phosphoglyceraldehyde. This is then converted to a hexose by the addition of one carbon to a five-carbon atom ring. In *Figure 5.23* an endeavour is made to summarize diagrammatically the salient points of the Calvin–Bassham cycle; this indicates the regeneration pathway of ribulose-1, 5-diphosphate. The overall scheme of photosynthesis is shown diagrammatically in *Figure 5.22*.

Thus the second phase of photosynthesis is the incorporation of atmospheric carbon dioxide; the entire process is extremely complex, and in fact not yet fully documented. However, it should be appreciated that fundamentally the carbon

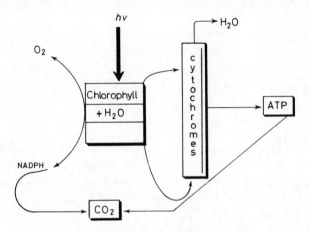

Figure 5.22. Overall scheme of photosynthesis

cycle is both a regenerative and a synthetic cycle. With regard to the regenerative aspect, reference is made to the point that when carbon dioxide becomes attached to the diphosphate ester of the pentose sugar ribulose, it produces a six-carbon molecule which is unstable, and thus subsequently breaks down into two three-carbon molecules of phosphoglyceric acid (PGA); then by a series of elaborate rearrangements, these triose phsophates regroup to form 1, 5-diphosphate. By such means the 'fixation' of atmospheric carbon dioxide is effected, and the carbon cycle is perpetuated. With the synthesis pathway, on the other hand, of those PGA molecules formed, five out of every six are used for the first reconversion pathway just described; the sixth PGA molecule can either be fed directly into the metabolic pool to yield ATP, for instance, or alternatively be upgraded in the synthesis of a higher polysaccharide molecule such as starch. Thus for every six turns of the Calvin-Bassham cycle one hexose molecule can be formed (*Figure 5.23*).

The principle of photosynthesis is the conversion of light into chemical energy. When electrons are excited they are raised to a higher energy level (orbital), and then returned to the ground state and photons may be emitted; such a situation forms the basis of fluorescence. The emission of light by organisms is often referred to as bioluminescence, and in nature there are a wide variety of such systems — for example, in *Photobacteria phosphoporeus*, Lampyridae (the fire flies), protozoa, radiolarans, and fungi. This ability to fluoresce is dependent on the presence of a substance called luciferin, of which there are many varieties. In the Lampyridae, the luciferin is known as 2-(4′-carboxythiazyl)-6-hydroxybenzo-thiazole, which reacts with ATP in the presence of a luciferase to produce fluorescence (McElroy and Seliger, 1962).

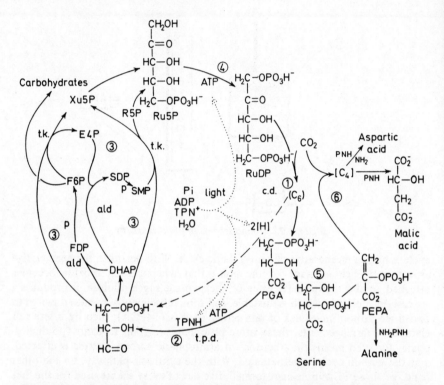

Key: c.d: carboxydismutase; PGA: 3-phosphoglyceric acid; tpd: triose phosphate dehydrogenase; DHAP: dihydroxyacetone phosphate; FDP: fructose-1,6-diphosphate; F6P: fructose-6-phosphate; tk: transketolase; E4P: erythrose-4-phosphate; 5DP: sedaheptulase-1,7-diphosphate; SMP: sedaheptulase-7-phosphate; Xu5P: xylulose-5-phosphate; R5P: ribose-5-phosphate; Ru5P: ribulose-5-phosphate; R1, 5DiP: ribulose-1, 5-diphosphate

Figure 5.23. The Calvin-Bassham Carbon Reduction Cycle (simplified): (1) ribulose diphosphate reacts with carbon dioxide to an unstable 6-carbon compound which splits to yield two 3-carbon compounds; (2) PGA-3-phosphoglyceric acid is reduced to 3-phosphoglyceraldehyde with ATP and NADPH derived from the light reaction and water; (3) various condensations and rearrangements convert the triose of (2) to pentose phosphates; (4) pentose phosphate is phosphorylated with ATP to give ribulose diphosphate; (5) further carbon reductions via conversion of PGA to phosphoenol-pyruvic acid; and (6) carboxylation, to form a 4-carbon compound (from Calvin, 1962; reproduced by courtesy of the author and Academic Press)

MUSCULAR CONTRACTION

Cilia and Flagellar Motion

Cilia are motility processes which are to be found in animal cells and some plant cells. These contractile filaments are called cilia if short and numerous, and flagella if long and sparse. They may undergo modification depending on their locale in a particular species. For instance, many bacteria possess flagella, and in the protozoa one class is characterized by their presence, the Flagellata. Similarly some of the Platyhelminthes, for example, possess external vibratile cilia. In higher animals, as in the mammalian system, ciliated epithelial cells are to be found lining the internal cavities and passageways of the respiratory tract and areas of the genital tract also.

Basically the ciliary apparatus consists of the cilium, a cylindrical process which originates from a basal body embedded in the cell soma — the ectoplasmic layer beneath the cell surface. The ultrastructures of the flagellum and cilia were found to be of a fundamental construction (*Figure 5.24*). In cross-section the total diameter of 2 000 Å constitutes an outer membrane surrounding the ciliary matrix. Embedded in this matrix are nine pairs of filaments, each about

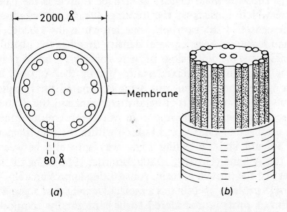

Figure 5.24. Diagrams showing general disposition of fibrils in a typical cilium. (a) A diagrammatic plan view — giving approximate dimensions; (b) three-dimensional structural view

80 Å in diameter, and there are two single filaments at the centre. The inner fibres are apparently conductors of stimulus, whereas the outer fibrils are contractors. Bradfield (1955), proposed that the flagella beat arises rhythmically from the basal body and spreads sequentially to the radial contraction fibrils. Shortening of

fibrils in one hemisphere causes the cilia to bend forward; subsequent relaxation in this area is followed by the initiation of contraction in the adjacent hemisphere, resulting in the recovery stroke. It has been found also that flagella hydrolyse ATP to ADP and phosphate, and during this chemical degradation ciliary motion is induced. According to Sleigh (1961) the metachromal rhythm of the ciliary beat is determined by the rate of contraction and excitation of a pacemaker cilium.

Muscle

Muscle represents what might be termed the 'equinox' of cell contractility development; structural organization has been adapted to unidirectional shortening during muscular contraction. Morphologically muscle cells are elongated and spindle shaped — the matrix being differentiated. The major portion of the muscle cytoplasm is occupied by contractile fibrils. In smooth muscle the myofibrils are homogeneous and birefringent. By contrast, in phase-contrast microscopy or in a stained preparation, cardiac and skeletal muscle reveal lateral striations consisting of dark birefringent anisotropic zones alternating with clear isotropic zones or bands (Hodge, Huxley and Spiro, 1954; Huxley, 1957) (*Figure 5.25*). Thus the main feature of striated muscle is the basic unit called the sarcomere which consists of the limiting Z band or telophragma which is located in the centre of the isotropic zone known as the I band. The A band is anisotropic and is seen to have a greater density than has the I band.

In vertebrate tissue the dimensions of these zones vary, but A is approximately 1.5μ and I is 0.8μ, in the relaxed state. Within the A band, under relaxed conditions, a lighter zone is seen, this being designated the H band or Hensen's disc. When muscle contracts this H band disappears and the Z and A bands are brought closer together. Furthermore in vertebrate muscle the constituent filaments are arranged in a hexagonal fashion within the thin filaments stretching through the I band and projecting some way into and between the thicker filaments of the A band (Huxley, 1960b; Bourne, 1960). The thick fibres of the A zone consist of the protein myosin, constituting approximately 75 per cent of the total muscle protein. Myosin has a molecular weight of some 450 000 and is an α-type fibrous protein, considered to be structurally composed of two or possibly three α-helices complexes, and along this myosin filament complex are numerous cross-striations showing a preiodicity of some 4 000Å (Hodge, 1956; Huxley, 1957) which are now referred to as interfilament bridges. Myosin subjected to digestion with trypsin can be split into two fractions called meromyosins—designated L and H, indicating light and heavy fractions respectively (Szent-Gyorgyi, 1953). The other major structural protein of muscle is actin, which accounts for approximately 25 per cent of the sarcoplasma protein. Actin has a molecular weight in the region of 70 000. In the absence of salts, actin exists as a globular protein (G-actin), whereas in the presence of potassium

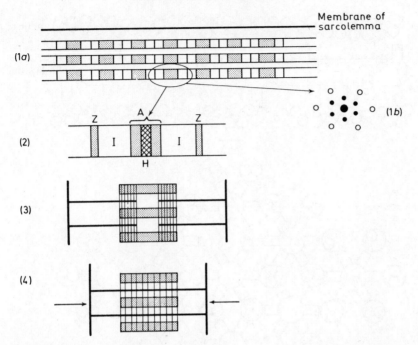

Figure 5.25. The basic unit of a striated muscle cell: (1) and (2). (1a) Depicts deposition of the thick and thin filaments within the sarcomere; (1b) cross-section of the A band. (2) Shows the sarcomere as seen under polarized conditions. The I and A bands are optically isotropic and anisotropic respectively. From Z to Z is referred to as one sarcomere. (3) and (4) Schematic diagram illustrating filament arrangement in striated muscle. The thick bar represents myosin in A band, and the thin bars extending from the vertical Z band are the actin filaments. (3) Represents a resting muscle configuration. (4) Indicates contracted muscle – the H band has disappeared and the Z bands have moved inwards towards the stationery A band

chloride and ATP it forms a long fibrous molecule (F-actin). This ability to change configuration is reversible, and is known as globular-fibrillar transformation. Actin forms the thin fibres of the Z band. Together *in vitro* these two sarcoplasmic proteins form actomyosin which in the presence of ATP and magnesium, contract (Szent-Gyorgyi, 1951). During this contraction ATP is degraded to ADP and inorganic phosphate. Hence the energetics and mechanics of muscle contraction are dependent fundamentally on the spatial relationship of actin and myosin filaments, together with the establishment of 'fuel' pathways, which are resident in the sarcoplasmic mitochondria, necessary to drive the muscular machinery. Phase-contrast microscopy has revealed that on contraction the

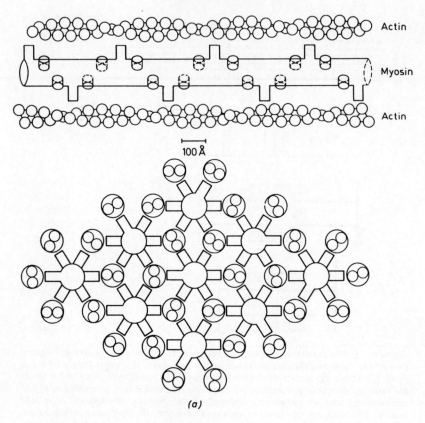

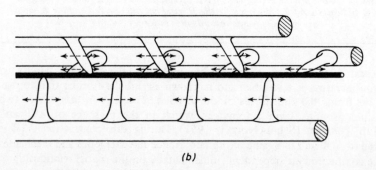

Figure 5.26. (a) The spatial arrangement of myosin and actin filaments in striated muscle (from Davies, 1963; reproduced by courtesy of the author and Nature, London). (b) Diagram indicating a possible mechanism of sliding between thick and thin filaments

width of the A band remains constant while the width of the I band changes accordingly, as the Z band is brought into close apposition with A. These microscopical observations have given rise to the sliding filament theory of muscular contraction (Huxley and Nierdergerker, 1954). In *Figure 5.26* the spatial relationship of myosin and actin filaments in striated muscle is depicted. It is assumed that the terminal ends of each projection in the myosin fibre possess enzymatic sites capable of activating degeneration of ATP. Hence contraction accentuates inter-digitation of the actin and myosin fibres (*Figure 5.25*), and subsequent degradation of ATP at the myosin active sites (Weber, 1960). But this hypothesis is not universally accepted for a problem still to be resolved is that of the isometric shortening of muscle.

ATP is the obvious immediate energy source in muscle activity; however, it has been shown that creatine phosphate acts as a reservoir of high-energy phosphate which can be re-phosphorylated to ATP. Creatine is synthesized in the liver from amino acids, and the major significance of creatine phosphate is that it is formed initially by the reaction of creatine with ATP. Hence, energy acquired from carbohydrate dissimilation in the form of ATP, and not utilized in muscle activity, is stored in the form of phosphocreatine:

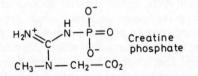

Creatine
phosphate

Similarly oxidation phosphorylation of ADP via the Krebs cycle is probably the most important source of energy – that is, ATP. Relaxation of contracted muscle is probably motivated by the conversion of an ADP actomyosin complex to an ATP actomyosin complex.

ACTIVE TRANSPORT

The passage of substances across a membrane in addition to simple diffusion is achieved by a mechanism described as active transport. Fundamentally the diffusion of a solute across a permeable membrane is a molecular flow from a high to a low concentration, this being motivated by virtue of a concentration gradient. Osmotic pressure can be shown to be directly proportional to the lowering of vapour pressure – which is dependent in turn on molar concentration. Thus experimentally the osmosis phenomena can be shown to be governed by laws analogous to the gas laws:

$$\Pi V = nRT$$

where Π is the osmotic pressure measured in atmospheres; V is the volume of the solution in litres; n the number of mols of solute present; T the temperature in degrees absolute; and R is the gas constant.

A molecular movement in the reverse direction — that is, against the concentration gradient — takes place by an active transport mechanism which is seen to require a supply of energy donated from the cell's metabolic activity, generally supplied in the form of ATP. This translocation of a substance against the concentration gradient is thus a promotion to a higher thermodynamic potential energy status, and with this gain there is seen to be a reciprocating loss of energy elsewhere in the system.

The amount of work (ΔG°) performed by a cell translocating a substance against the concentration gradient is:

$$\Delta G^\circ = (RT.\ln)\,([C_1]/[C_2])$$

where C_1 is the higher concentration, and C_2 is the lower concentration, these being the thermodynamic concentrations of the solute in each compartment.

The most convenient example of active transport is to be found in the erythrocyte, though there are numerous examples in cell complexes of the multicellular organism such as the parietal cells of the gastric mucosa, the mitochondrion, the nerve axon, in which ionic transportation is utilized, resulting in a transformation of chemical energy into either osmotic energy or electrical energy. The later transformation will be discussed in Chapter 6.

It had been found that there exists a differential concentration gradient between potassium and sodium ions connected with the red cell, there being a higher concentration of intracellular potassium ions and a correspondingly low concentration of sodium ions — as opposed to an extracellular concentration reversal of this situation. This ionic distribution in relation to the erythrocyte is the result of active transport mechanisms, in which energy derived from the glycolytic conversion of glucose yielding two molecules of ATP is utilized. If there should be a shut-down in the glycolytic pathway due, for example, to the inhibition of enzyme enolase participation by the chemical sodium fluoride, then the sodium: potassium ratio will be upset, since the two ions will equilibrate with the intracellular and extracellular environment. It has been deduced from this that active transport is an energy-dependent process. In the same way, in other cells, such as the mitochondrion, this energy dependence can be proved by the application of respiratory enzyme inhibitors.

Fundamentally, active transport serves three major interdependent functions required to maintain in turn cellular interdependent functions essential to the preservation of cellular integrity. These are: (1) control of pH; (2) maintenance of specific differential ionic concentrations; (3) maintenance of normal ordered enzyme activity by both (1) and (2).

The theoretical basis of active transport is summarized in *Figure 5.27* which depicts the cellular transfer of sodium and potassium by simple and active means. Simple transport is distinguished from active by considering the former as being 'down hill' as distinct from the latter which may be termed 'uphill'. Some theoretical models set up in an attempt to explain transport mechanics require that the solute being translocated be linked to a particular component within the membrane fabric to which a specific directional enzyme motivates re-orientation of the transported substance. Pursuing this concept workers supporting

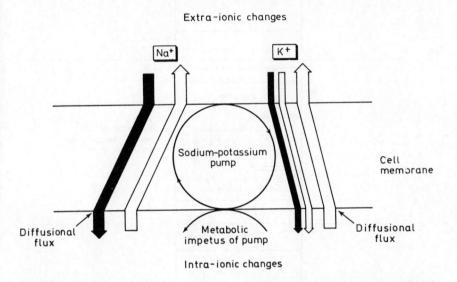

Figure 5.27. Schematic depiction of active and passive sodium and potassium transport. The width of each arrow indicates the size of that particular one-way flux. As the diagram implies the passive efflux of sodium is negligible − and hence is not shown

such a theory have shown that when an ATPase enzyme molecule hydrolyses intracellular ATP, hydroxyl ions are utilized from the opposite direction and, as a consequence, ADP and $^-HPO_4$ become products (*Figure 5.28*) (Whittam, 1962). Furthermore ATPase activity and ATO degeneration have been shown to be dependent on the spatial differences in sodium and potassium concentrations existing on both sides of a membrane. The precise details of this model have not been determined, but it is thought to be reliant on the existence of vectorial enzymes resident in the membrane structure. The foregone concept is summarized in *Figure 5.28*. For alternative theories *see* Bibliography (Ussing, 1957).

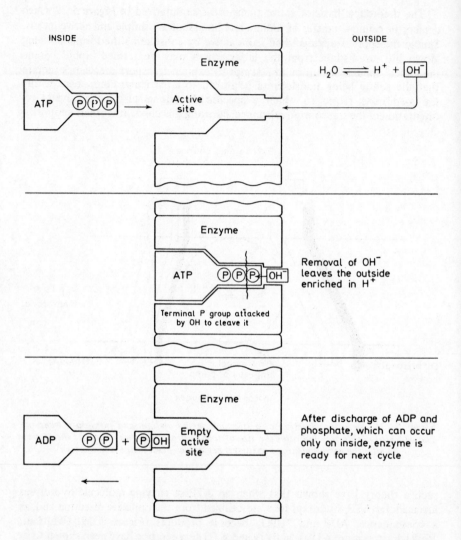

Figure 5.28. Asymmetric ATPase molecule in a membrane can cause removal of OH⁻ ions, from the outside during the hydrolysis of ATP to ADP 4 P in the inside compartment. Since the product of H^+ and OH^- is always constant (1×10^{-14} at $25° C$), depletion of OH is accompanied by a rise in H in the outer compartment (from Lehninger, 1965 in Bioenergetics: The Molecular Basis of Biological Energy Transformations: reproduced by courtesy of Benjamin Inc., New York)

ACTION POTENTIALS

The basis unit of the central nervous system is the neurone which consists of the perikaryon, from which the dendrites arise, constituting the dendritic zone. Projecting from this cell body is a single fibrous rod called the axon which carries the nerve impulse. The axon is composed of a lipid protein complex in association with Schwann cells forming an exterior protective layer called the neurilemma. The axon length is punctuated at regular intervals to give constrictions or nodes of Ranvier which are regions of low insulating resistance. There is usually one Schwann cell nucleus present per internodal segment of myelin. The dendritic zone, activated by a stimulus exceeding the excitation threshold, transmits an impulse along the axon rod which terminates ramifying in the axon telodendrion — the location site of the synapse. The synapse is a bulbous expansion occurring at the axon terminal, known as the pre-synaptic bulb, which is separated from the post-synaptic membrane of an adjacent dendrite by a space called the synaptic cleft. Hence this section is concerned superficially with the conversion of varied stimuli into electrical energy and its subsequent propagation along the axon rod, and the transmission of this impulse across the synaptic cleft.

Essentially nerve cells have undergone specialization whereby the active transport of sodium and potassium ions across the cell membrane is exploited in the propagation of a nerve impulse. This is achieved by virtue of changes occurring in the resting membrane potential. Such a cell at rest possesses a stable electrical potential status which is maintained by active transport and passive diffusion — such that there exist differential concentration gradients—that is, where the internal potassium concentration is greater than the external concentration; the reverse obtains in the case of sodium ions. Hence the interior of the cell compartment is negative, whereas the exterior is positive. The magnitude of this potential difference is seen to be a function of ionic distribution about the axon membrane. Thus the sodium pump mechanism serves to maintain the interior negative status of the membrane. As a result of excitation there is an increase in local permeability of the axon membrane resulting in a redistribution of sodium and potassium ions across the membrane precipitating in a depolarized region.

This active depolarized region motivates a directional current flow between it and the adjacent polarized region. The normal electrical potential difference following an action potential is rapidly restored by the mediation of ionic pumps which require the participation of ATP. Thus an action potential possesses the following characteristics: (1) all-or-none response: that is, specific neurones will only become activated by a stimulus or impulse of sufficient magnitude, and the minimum stimuli strength capable of activating a neurone is known as the threshold intensity. Any further increase in intensity will not produce any further increment in activity. Thus the amplitude of the action potential remains the same once the threshold of activation for a particular type of neurone has been reached; (2) similarly, the amplitude of the action potential does not decrease along the axon length. Thus the nerve impulse is said to be non-decremental; (3) following the passage of an action potential over a particular area, there follows a refractory period, or period of recovery.

The transmission of an impulse across a synaptic or myoneural junction requires the participation of a transmitter substance which is thought to be resident in the synaptic vesicles. The ultrastructure of the synapse consists of mitochondria and numerous capsule-like bodies called vesicles, the approximate dimensions of which are 500 Å. The pre-synaptic membrane contains intra-projecting bundles of protein fibres, arranged like wigwams, which are thought to be vesicle receptor sites. It would appear that these vesicles contain transmitter substances of which acetylcholine is thought to be one; these are substances capable of exciting the receptor sites. With the arrival of an axonal impulse, this initiates the migration of these vesicles to the receptor sites on the pre-synaptic membrane. These orientated vesicles then release their excitatory substances which cross the synaptic cleft to the complementary receptor sites situated on the post-synaptic surface. With the existence of this transmitter substance acetylcholine there is obviously a mechanism at the pre-synaptic site for both the removal and regeneration of the transmitter substances. In fact there is a substance called acetylcholine-esterase. By virtue of its geometry this substance possesses two receptor sites to which acetylcholine can become anchored. Furthermore acetylcholine can be synthesized from active acetate and choline. Also studies have revealed that acetylcholine-esterase is present in high concentrations at the axon terminals. Thus in a normal resting nerve acetylcholine is bound to a storage protein in the pre-synaptic vesicle; this, following the arrival of an impulse, is released. The free acetylcholine can then: (1) be degradated by acetylcholine-esterase; or (2) become attached in a reversible manner to a receptor protein molecule; (3) binding with this receptor protein precipitates conformation changes in the protein structure which in some way results in local synaptic membrane depolarization; (4) equilibrium then reverses the direction of the choline and acetate, which seduces the acetylcholine away from the receptor sites, thus restoring it to its original configuration, and the membrane to its normal bioelectrical status. However, from this it will be seen that synaptic transmission of a nerve impulse has still to be clarified.

BIBLIOGRAPHY AND REFERENCES

Anderson, B. and Ussing, H. H. (1960). 'Active Transport.' *Comp. Biochem. Physiol.*, 2, 371.
Arnon, D. I. (1955). 'The Chloroplast as a Complete Photosynthetic Unit.' *Science, N.Y.*, 122, 9.

Bassham, J. A., Benson, A. A., Kay, L. D., Harris, A., Wilson, A. T. and Calvin, M. (1954). 'The Path of Carbon in Photosynthesis.' *J. Amer. chem. Soc.*, **76**, 1760.

Bourne, G. H. (1960). *Structure and Function of Muscle.* New York; Academic Press.

Bradfield, J. R. G. (1955). 'Fibre Patterns in Animal Flagella and Cilia.' *Symp. Soc. exp. Biol.*, **9**, 306.

Calvin, M. (1962). 'Evolutionary Possibilities for Photosynthesis and Quantum Conversion.' In *Horizons in Biochemistry*, pages 23–26. Ed. by Kasha, M. and Pullman, B. New York; Academic Press.

– and Bassham, J. A. (1962). *Photosynthesis of Carbon Compounds.* New York; Benjamin.

Chance, B. (1956). 'Interaction of Adenosine Phosphate with the Respiratory Chain.' International Symposium, 447, Henry Ford Hospital.

– (1963). *Energy Linked Functions of Mitochondria.* New York; Academic Press.

Davies, R. E. (1963). 'A Molecular Theory of Muscle Contraction.' *Nature, Lond.*, **199**, 1068.

Eccles, J. C. (1957). *The Physiology of Nerve Cells.* Baltimore; Johns Hopkins Press.

Green, D. E. (1956). 'Structural and Enzymatic Pattern of the Electron Transfer System.' International Symposium, 465, Henry Ford Hospital.

– (1962). 'Structure and Function of Subcellular Particles.' *Comp. Biochem. Physiol.*, **4**, 81.

– and Fleischer, S. (1960). *Metabolic Pathways*, vol. 1. New York; Academic Press.

Hill, R. (1939). 'O_2 Produced by Isolated Chloroplasts.' *Proc. R. Soc. B.*, **127**, 192.

– (1940). 'Production of O_2 by Illuminated Chloroplasts.' *Nature, Lond.*, **146**, 61.

Hodge, A. J. (1956). 'The Fine Structure of Striated Muscle.' *J. biophys. biochem. Cytol.* 2. Suppl. 131.

– Huxley, H. E. and Spiro, D. (1954). 'Electron Microscope Studies of Ultra-thin Sections of Muscle.' *J. exp. Med.*, **99**, 201.

Huxley, A. F. (1957). 'Muscle Structure and Theories of Contraction.' *Prog. Biophys. biophys. Chem.*, **7**, 255.

– and Nierdergerker, A. (1954). 'Structural Changes in Muscle during Contraction.' *Nature, Lond.*, **173**, 971.

Huxley, H. E. (1960a). 'The Contraction of Muscle.' *Scient. Am.*, **199**, 67.

– (1960b). 'Muscle Cells.' In: *The Cell*, vol 4, page 365. Ed. By J. Bracket and A. E. Mirsky. New York; Academic Press.

Jencks, W. P. (1962). *The Enzymes*, vol. 6, page 399. New York; Academic Press.

Krebs, H. A. (1950). 'The Tricarboxylic Acid Cycle.' *Harvey Lect.*, Ser. **44**, 165.

Kleinzeller, A. and Kotyk, A. (Eds.) (1961). *Membrane Transport and Metabolism.* New York; Academic Press.

Lehninger, A. L. (1961). 'How Cells Transform Energy.' *Scient. Am.*, **205**, 62.

– (1965). *Bioenergetics: The Molecular Basis of Biological Energy Transformations.* New York; Benjamin.

McElroy, W. D. and Seliger, H. H. (1962). 'Mechanisms of Action of Firefly Luciferase.' *Fedn. Proc.,* **21**, 1006.

Masoro, E. J. (1968). *Physiological Chemistry of Lipids in Mammals.* Philadelphia, Pa.; Saunders.

Rustad, R. C. (1961). 'Pinocytosis.' *Scient. Am.,* **204**; 121.

Singer, T. P. (Ed.) (1966). *Biological Oxidations.* New York; Wiley.

Sleigh, M. A. (1961). 'An Example of Mechanical Coordination of Cilia.' *Nature, Lond.,* **191**, 931.

Szent-Gyorgyi, A. G. (1951). *Chemistry of Muscular Contraction,* 2nd ed. New York; Academic Press.

— (1953). 'Meromyosins, The Sub-units of Myosin.' *Archs. Biochem.,* **42**, 305.

Ussing, H. H. (1957). 'General Principles and Theories of Membrane Transport.' In *Metabolic Aspects of Transport Across Cell Membrane.* Ed. by R. R. Murphy. Madison, Wis.; University of Wisconsin Press.

Weber, H. H. (1960). *Molecular Biology.* Ed. by D. Nachmansohn. New York; Academic Press.

Whittam, R. (1962). 'Asymmetrical Stimulation of a Membrane Adenosine Triphosphatase in Relation to Active Cation Transport.' *Biochem. J.,* **84**, 110.

Part IV
The Basis of Somatic Integrity

6 — Basic Genetics

(Cell division and growth − Genes and chromosomes − The genetic code − Protein synthesis − Control of protein synthesis − Classification and recognition of normal chromosomes − Sex and autosomal chromosomes − Inborn errors of metabolism)

A GENERAL CONSIDERATION OF CELL DIVISION AND GROWTH

One of the most fundamental characteristics of living cells is their property of growth, which can be defined as an orderly increase in all the components constituting the organization of that organism. During this process exchanges between the cell and its environment occur − namely, the absorption of food molecules and their transformation into cellular components such that the organism will exhibit a positive addition to its existing body material, and anabolism is seen to exceed catabolism.

The whole subject of cell growth is multifactorial and will not be considered here in depth since it extends beyond the intended subject content of this section. The reader is advised to consult more definitive literature to enlarge upon this (*see* Bibliography).

The normal cycle of a cell includes one complete interphase period and one cell division, and the events of this cycle represent an increase in cell size, replication of the genetic code, division of the cell mass, and the subsequent duplication of the genetic code in two independent functional units. Hence in a multicellular organism multiplication ultimately leads to an increase in size, whereas in the unicellular organism there is an increase in the number of separate individuals, since, − following cell division − there is a mutual separation between units: in other words they inherently possess an anti-aggregation mechanism. Cells of the multicellular organism, on the other hand, exhibit a reverse tendency, in as much as they remain in close apposition to one another, forming a harmonious aggregate − all of which serves to indicate an evolutionary progression towards increased mutual adhesion. Furthermore it would seem that in the higher animals the rate of cell division is closely

171

related to the pattern of cell growth, and it is conjectured that both share the same regulatory mechanism.

Prior to discussing mitotic division of a cell, cellular growth will be briefly considered. Under ideal environmental conditions, reproduction and growth of a cell will occur according to a geometric progression with time. This is probably best illustrated using a bacterial model, such as a culture of *Escherichia coli*. Here each bacterium grows and, on reaching a particular size, will divide by binary fission thereby producing two individuals which will possess physiological characteristics similar to those of the parent cell. Thus they will grow and then divide again at approximately the same rate. Initially in such a culture the first generation divisions will be synchronized but after an interval of time, because of differences occurring among the new population, the time of division will become randomized. In such a culture the number of cells double with each generation, thus the population is seen to increase as the power of 2, and the log 2 of the number of organisms will increase in direct proportion with time. This is termed the principle of exponential growth (*Figure 6.1(1)*).

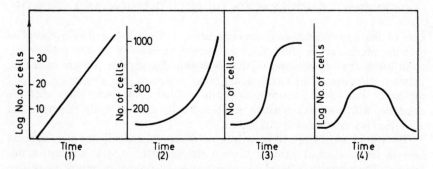

Figure 6.1. Graphical depiction of various growth curves: (1) and (2) represent exponential curves; (3) sigmoid curve, representing growth of any population under natural conditions; (4) typical growth curve of increase of cell numbers as a function of time

Thus, by virtue of the exponential nature of growth of the unicellular organism, the growth rate is expressed in terms of the generation time – which is that period of time required for a growing population to double itself – rather than in terms of an increase in the number of cells per unit of time. Since such an increase with each generation is represented by one exponent of 2, the number of generations is equal to the difference between the two exponents of 2, corresponding to the two numbers. Thus the number of generations may be calculated as:

$$\text{Log}_2 \ N_1 - \text{Log}_2 \ N_o$$

where N_o is the initial population number and N_I is the population after a particular interval of time. Or more conveniently it may be expressed using ordinary logarithms:

$$(\text{Log}_{10} N_1 - \text{Log}_{10} N_0)/(\text{Log } 2) = (\text{Log}_{10} N_1 - \text{Log}_{10} N_0)/(0 \cdot 301)$$

In *Figure 6.1*, four graphs of growth curves are depicted: (1) and (2) represent exponential curves which mathematically may be expressed thus:

$$Nt_2 = Nt_1 e^{kT}$$

where Nt_2 represents total number of a population at any time t_2 — that is, the time interval over which calculation is made; Nt_1 is total number present at the commencement of the time interval over which calculation is made; e is log base $2 \cdot 718$; k is a constant representing the efficiency of the population to increase its numbers, and T is overall duration of experimental time.

If the above is applied (namely, *(1)* and *(2)* in *Figure 6.1*) to a multicellular organism then theoretically the population would be extremely large. Indeed, a single bacterium having a generation time of 20 minutes would produce in 48 hours of exponential growth $2 \cdot 2 \times 10^{43}$ cells — and furthermore if the average weight of a single bacterium was taken as 10^{-12} g, then the total weight of this 48-hour culture would be astonishingly great, $2 \cdot 2 \times 10^{31}$ g. But in reality, as in *in vitro* culture, this reproduction potential is never achieved, since certain limiting factors come into play, which retard and eventually terminate the growth rate. *Figure 6.1 (3)* is a sigmoid curve which in fact represents the growth of any population under natural conditions. On the other hand *Figure 6.1 (4)* is a typical growth curve which describes the increase of cell numbers as a function of time. Thus in the case of (2) under both natural and experimental conditions the bacterial population becomes limited by the eventual exhaustion of nutrients or by the accumulation of toxic substances. That is, the products of fermentation will eventually interfere with the normal metabolism, with the net result that the growth rate will rapidly decrease to zero. Thus graph (4) represents an historical record of the interaction of a growing cell population and its environment.

As implied earlier, synchrony of cell division rapidly becomes randomized, and yet if cells are required, for example, in mitotic studies random division is a severe technical handicap. One way of achieving synchronous cultures is to utilize the so-called temperature shock method, which first requires inhibition of cell division by incubating the culture at sub-optimal growth temperatures and then subjecting the culture to a sudden temperature change which will encourage mitosis. Exploitation of this, and other techniques such as filtration and treatment with amethopterin will, it is hoped, eventually lead to more definitive knowledge of those mechanisms regulating the events of the cell cycle (Zeuthen, 1964; Brachet and Mirsky, 1959).

A Description of Mitotic Division

In *Figure 6.2* the general scheme of mitotic division of a cell is shown diagrammatically including early and late phases. This cycle can be divided for convenience into four main stages in which the nucleus undergoes a continuous series of complex re-organization manoeuvres. The interphase nucleus at first appears featureless — with the exception that in some cells the nucleus may possess visible nucleoli.

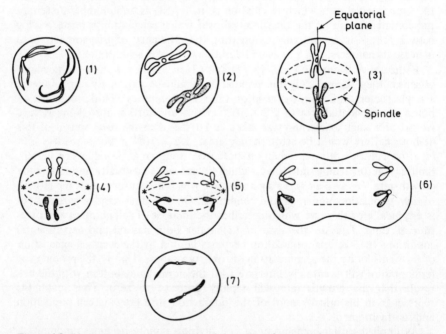

Figure 6.2. Diagrammatic representation of the salient features of mitosis in a haploid cell. (1) Early prophase, in which each chromosome is seen as two sister chromatids. (2) Prophase period — during this stage the chromatids become localized in a region called the centromere, and as a consequence are reduced and thickened in size. (3) Metaphase period: this heralds the formation of spindle apparatus across the nuclear area along which the chromosomes become orientated in an equatorial position. (4) Late metaphase: division of each chromosomal centromere. (5) Anaphase: migration of daughter chromosomes to opposite poles. (6) Telophase: re-enactment of prophase in reverse, together with cytokinesis. (7) Formation of nuclear membrane and the re-establishment of nuclear interphase conditions

Prophase

The first stage of mitosis is prophase, and this initial activity is characterized by the loss of nuclear homogeneity, and the appearance of delicate longitudinally coiled threads within the nuclear matrix — these being the 'embyronic' chromosomes. Each chromosome is composed of two coiled filaments collectively known as the chromatids, and separately as the chromonemata.

During this initial differentiation the chromatids appear to be localized by a zone called the centromere which seems to mediate in a mechanism controlling the correct spatial orientation of the chromosomes to the cell space. Occurring simultaneously with this is the formation of the spindle, the metamorphosis of which originates from the appearance of the extra-nuclear centrioles. Each centriole produces an aster, with radial striations known as the astral ray. Hence as the prophase develops, each centriole, consisting of two bodies respectively, migrates — eventually becoming aligned at opposite ends of the nuclear membrane. Arising between the juxtaposed asters is a stream of delicate threads, which form a spindle. The manner in which the spindle is formed in conjunction with centriole migration is very much dependent on the cell type in question. It has been shown that the spindle fibrils stop short of the centriole region. Furthermore the coarse structure of the centriole is known to be a cylinder approximately 150 mμ in diameter, containing small tubules of approximately 200 Å diameter, which are orientated parallel to the axis. Generally the centriole is found in the interphase cell, in association with other cytoplasmic organelles, such as the Golgi apparatus.

In the closing stages of prophase — sometimes referred to as prometaphase — each chromosome appears to be two distinct longitudinal elements which become positioned on the nuclear boundary. At the same time the nucleus becomes elongated along its polar axis, which eventually ruptures and disappears.

Metaphase

Thus final disintegration of the nuclear membrane occurs in metaphase, together with the extension of the spindle fibrils across the central area of the cell. With this achieved the spindle fibrils apparently attract the chromosomes to assemble themselves equatorially in what appears to be a specific spatial sequence on the spindle fabric — each attached to a spindle fibril by its centromere. Some workers believe that the spindle fibrils originate from the centromere, and grow towards the polar-orientated centrioles, but this is still largely supposition.

Anaphase

The next fundamental change occurs in anaphase in which each centromere and chromatid divides. The resultant daughter centromeres now move away from each other pulling the chromatids, now called daughter chromosomes, apart; they then migrate towards a particular cell pole.

Telophase

From now on there are two daughter chromosomes per centromere, rather than a single chromosome composed of two chromatids held by a single centromere. This polar migration is known as telophase and is the final stage of mitosis, and may be visualized as prophase in reverse. This migratory movement of the chromosomes to the cell poles, is believed to be due to contraction of the spindle fibrils activated by ATP. During the latter stages of telophase the chromosome complex becomes less compact by an apparent relaxation, which motivates an uncoiling of the tightly bound chromonemata. Eventually these structures become enclosed by a membranous material called the karyoplasm. At the same time reconstruction of the nuclear membrane occurs. Telophase terminates with cytokinesis, during which cytoplasmic cleavage occurs. It is important to stress here that the actual mechanics of cytokinesis, as indeed of all stages of the mitotic cycle described, are quite unknown (Mazia, 1961a).

TABLE 6.1

Diploid (2n) Number of Chromosomes in some Plants and Animals

Animal or plant species	Chromosome number
Homo sapiens	46
Macaca mulatta	48
Felis maniculata	38
Rattus rattus	42
Gallus domesticus	78
Rana esculenta	26
Musca domestica	12
Culex pipiens	6
Brassica oleracea	18
Malus sylvestris	34 and 51
Lathyrus odoratus	14
Phaseolus valgaris	22
Solanum tuberosum	48
Secale cereale	14

A Description of Meiotic Division

The somatic cells of plants and animals possess a normal complement of two pairs of chromosomes. An individual member of a chromosome pair is called a homologue. In man the normal complement is 23 pairs or 46 homologues. Table 6.1 shows the diploid numbers of chromosomes in a variety of animal and plant species.

Whatever the normal complement for a particular species, the original chromosome number, the diploid number (2n), is preserved as a result of nuclear division occurring during the development of a multicellular organism. Hence mitosis is a mechanism which ensures the conservation of genetical material within each daughter cell. Meiosis, on the other hand, achieves reduction of the chromosome complement, from the diploid number (2n) to the haploid number (n). Thus organisms reproducing by means of a sexual mechanism — that is, by utilizing the union of two gametes, which if diploid would result in the zygote having twice the normal diploid chromosome number — are maintained constant by meiosis. Consequently, meiotic division of each gamete reduces the normal diploid set to a single haploid number — all of which occurs during the course of normal gametogenesis. The reduction of the chromosome number is achieved by two nuclear divisions — namely, the first and second meiotic divisions. In the first of these the homologous chromosomes form pairs and then segregate into daughter cells; in the second stage schism of each homologue occurs, the products of which subsequently enter into each of the four resultant cells. For a diagrammatic summary of meiosis *see Figure 6.3.*

Meiotic Division I

Meiosis commences with a long first prophase, which for convenience is divided into four intermediate stages: leptonema, zygonema, pachynema and diplonema. In the leptonema stage the nuclear area becomes differentiated, with the appearance of the chromosomes as long discrete threads. The zygonema stage is characterized by the pairing of the homologous chromosome threads, this pairing apparently being specific. With pairing completed, the nucleus enters the pachynema stage in which the chromosomes contract longitudinally into shorter compact structures. Each chromosome possesses an independent centromere and — as a consequence of longitudinal cleavage during this period — each pachytene consists of four chromatids. Simultaneous transverse fission may occur at identical points on two homologous chromatids which results in an exchange of segments between the chromatid pair. Finally at the diplonema stage, separation of the chromosome population occurs. This repulsion is seen to be restrained by the presence of the chiasmata, which are the points of interchange described in the late pachynema period. These chiasmata are regarded as an expression of the 'crossing-over' phenomenon which will be discussed later. The later stages of this long initial prophase are characterized by chromosome contraction known as diakinesis, and migration of chiasmata towards the terminal end of the chromosome soma.

Metaphase I commences with the disintegration of the nuclear membrane and the positioning of the chromosomes along the cell equator, with their centromeres orientated towards opposite poles along a spindle which has formed. Anaphase I exhibits migration of the chromosomes to opposite cell poles, and during this stage the chromosome composition differs from that of the

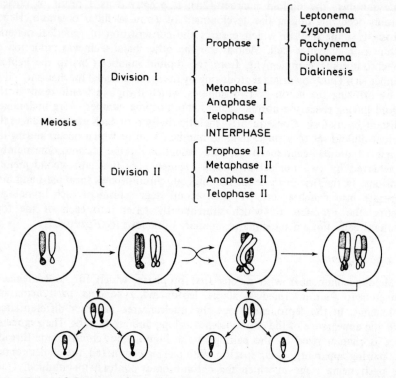

Figure 6.3. Summary of meiotic division – for simplicity only one chromosome pair is represented

original. Telophase I, starts with the arrival of the chromosome groups at the polar reception sites. Thus with the completion of the first telophase the net result is the halving of the numbers of centromeres per cell. The number of chromatids per cell following the first meiotic division is then equal to the diploid number (2n) in somatic cells.

Meiotic Division II

This second stage follows a short interphase period, commencing with an equally brief prophase II, in which the nuclear membrane disappears again, and the spindle is re-formed. Metaphase II and anaphase II are characterized by the lining up of the chromosomes along the equator of each cell and the division of the centromeres. The two resultant sister chromatids then migrate to opposite poles. With the completion of telophase II there are four haploid cells, in which each original chromosome is represented once.

Thus in meiotic division I, the chromatids remain linked to the centromere. In consequence, during metaphase I, when the separation occurs, each pair goes to the same daughter cell. In the second meiotic division the chromatids divide mitotically, and are thus distributed to each daughter cell. In addition to reducing the diploid number to the haploid number set, meiosis differs in two further major ways from mitosis, in exhibiting the segregation of the chromatids and the 'crossing-over' phenomenon.

Law of Segregation

The principle of Mendel's first law of genetics is summarized in *Figure 6.4.* When two contrasting varieties of plants are crossed by artificial pollination

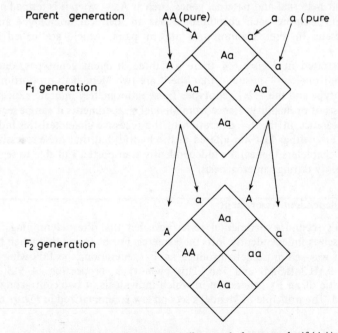

Figure 6.4. The law of segregation illustrated: for example, if (AA) represents a homozygous red flower, and (aa) a white flower, and the heterozygous combination is represented by the colour pink – then the 1:2:1 ratio would be red: pink/pink: white

179

the resultant progeny of the first generation (F_1), always resembles one of the initial parent types. However, on re-crossing F_1 individuals, those characters not seen in the F_1 generation subsequently re-appear in the new F_2 generation. This forms the basis of Mendel's first law; for from a large series of experiments utilizing the above concept Gregor Mendel deduced that those characteristics manifested in the first generation were 'dominant', whereas those which had disappeared were 'recessive'. Mendel found that these recessive characters re-emerged in the F_2 generation in the proportion of 3:1 — dominant: recessive. He concluded that generation characters are controlled by pairs of inherited factors (later called genes), one of which was dominant over the other; these are capable of being separated from one another during fertilization, and are able subsequently to recombine in definite proportions.

The morphological characteristics manifested by an organism are together referred to as the phenotype. The term genotype refers to the genetic constituent of an organism, which is the net result of all its hereditary potential. The resultant zygote is termed homozygous when it possesses two like factors (for example, AA), in contrast to one composed of two unlike factors — that is, different maternal and paternal genes, such as Aa — when it is termed heterozygous, Furthermore each chromosome has an identical homologue and the genes present in the chromosomes are in pairs, which are called allelic pairs.

As illustrated in *Figure 6.4* there are three different genotypes: one AA, two Aa and one aa — meaning that there are two Mendelian proportions, the 3:1 phenotype and the 1:2:1 genotype. These ratios merely indicate expectation on the basis of probability. Hence from Mendel's experiments it can be seen that the reappearance in the F_2 generation of the recessive characteristics indicates that they are neither lost nor altered in the hybrid, but that these recessive and dominant characters (genes) are independently transported and able to segregate independently during gametogenesis.

Law of Independent Assortment

Mendel's second experimental series revealed that the determining factors sort themselves independently into the gametes. For he found that when the F_1 generation was self-crossed to produce an F_2 generation, the following ratios emerged: 9AB,3aB,3AB and 1ab. This phenotypic proportion of 9:3:3:1 is characteristic of an F_2 generation in which individuals of two contrasting genes are crossed. The principle of Mendel's second law is summarized in *Figure 6.5*.

Linkage and Crossing-over

The studies of Morgan with Drosophila demonstrated that Mendel's second law was not universally applicable, and that for certain crosses of two or more alleles there was a certain limitation of segregation. The reason for the failure of two genes to assort independently is that they are located on the same chromosome — that is, they are linked. The frequency of segregation can be

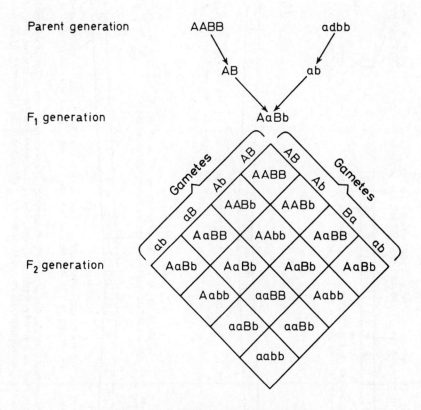

Figure 6.5. Law of independent assortment

altered by chromosome fusion or breakage, but probably the most important mechanism existing for the exchange of genes on homologous chromosomes is crossing-over, which occurs particularly during meiosis. For at the commencement of meiosis the homologous chromosomes form pairs, and each chromosome is duplicated to form two chromatids — and during the first meiotic division, probably between leptonema and pachynema or zygonema in diploid cells, an exchange takes place between the adjacent chromatids. With the establishment

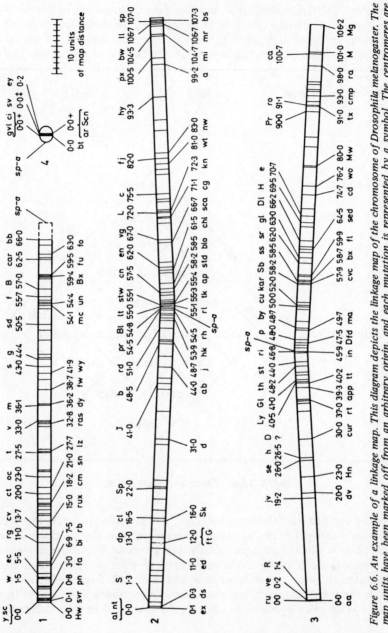

Figure 6.6. An example of a linkage map. This diagram depicts the linkage map of the chromosome of Drosophila melanogaster. The map units have been marked off from an arbitrary origin, and each mutation is represented by a symbol. The centromeres are represented by the symbol sp-a (from Sinnott, Dunn and Dobzhansky, 1958; reproduced by courtesy of McGraw-Hill)

of chiasmata at about the pachynema and diplonema stage, fission of two of the four chromatids and reunion occur at each chiasma such that the new chromatids are composed of fragmented sections of the original chromosome assembly. This re-organization of genetic material is called re-combination and, as implied above, the frequency of re-combination of two linked genes is a function of the distance separating them. It is important to note here that some workers (for example, Pritchard, 1960) consider pairing occurs during the pre-meiotic interphase when, it is believed, adhesion of homologous chromosomes occurs as a result of random movement of chromosome material in the nuclear region. Subsequently, the time of cross-over is obviously detectable if it had occurred by the close of pachynema and, in Pritchard's opinion, cross-over occurs at the time of early homologous contact, since even in the early stages of prophase I chromosome contact is already apparent.

As a result of Morgan's work it was found that genes are distributed in linear formation along a chromosome, and that the location of a particular gene is constant. Hence by means of recombination this can be used as a means of determining gene proximity within a group, and thus of assembling subsequently what is known as a chromosome map: *Figure 6.6* is an example of such a map. The distance between genes therefore is simply a unit of re-combination. Take, for instance, the simple gene complex A, B, and C. Supposing the distance between A and B was x, and B and C was y, then the distance between A and C is simply the sum of the approximate x and y units. In *Figure 6.6* a linkage map of *Drosophila melanogaster* is shown: it must be appreciated that these maps show only approximations of distances, and thus only represent a relative gene order. In Drosophila there are four pairs of chromosomes in a haploid cell, each possessing one group of genes. The genetic lengths of these four chromosomes are 66 units for the first, 107·3 units for the second, 106·2 for the third and 0·2 units for the fourth. In *Figure 6.6* the genes have been assigned to four distinct linkage groups by virtue of the fact that the end gene exhibits a 50 per cent re-combination with other gene groups, but tends to link with other genes of its own group.

As already stated, of each chiasma only two of the four chromatids cross over — such that, of the four chromosomes resulting from meiotic division, only two can exhibit this phenomenon at any one level. Proof of this crossing-over between two loci and involving only two of the four chromatids has been obtained from studies of Drosophila and the mould Neurospora. This concept is illustrated in *Figure 6.7*.

Gene Frequency

Consider now a population when at a particular locus on an autosomal chromosome gene A, or its allele gene a, may occur. According to Mendelian

183

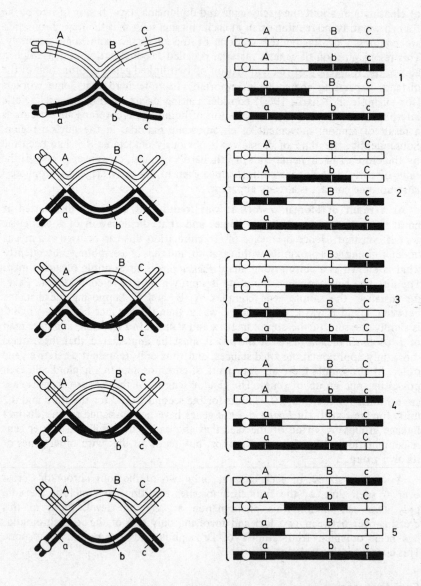

Figure 6.7. Diagram showing chromosome bivalents with single chiasmata (above) and double chiasmata (below). (1) Single chiasma; (2) two-stranded double cross-over; (3) four-stranded double cross-over; (4) and (5) three-stranded double cross-over

genetics here there are three genetically distinct types of individuals possible — AA, Aa and aa. If the number of individuals of the three genotypes are designated α, β and γ, then the frequency (p) of gene A in the population is:

$$p = (\alpha + \beta/2)(\alpha + \beta + \gamma)$$

and the frequency (q) of gene a in the population will be:

$$q = (\gamma + \beta/2)(\alpha + \beta + \gamma)$$

$$\text{so that: } p + q = 1$$

Thus the relative proportions of the three types of individual in a randomly mating population are: $AA = p^2$; $Aa = 2pq$; and $aa = q^2$

From this it can be seen that the gene frequency and the genotype frequency are the same in both the offspring and the parental population; this is often referred to as the Hardy—Weinberg equilibrium. Table 6.2 tabulates the frequencies of random parental mating types and their offspring for an autosomal gene pair.

TABLE 6.2

Parents		Offspring		
Mating types	Resultant frequency	AA	Aa	aa
AA $\times$ AA	p^4	p^4		
AA $\times$ Aa	$4p^3q$	$2p^3q$	$2p^3q$	
AA $\times$ aa	$2p^2q^2$		$2p^2q^2$	
Aa $\times$ Aa	$4p^2q^2$	p^2q^2	$2p^2q^2$	p^2q^2
Aa $\times$ aa	$4pq^3$		$2pq^3$	$2pq^3$
aa $\times$ aa	q^4			q^4
Total matings		p^2	$2pq$	q^2

Mutation and Aberration

Mutations are regarded as the ultimate source of new and different material appearing in a population; whereas re-combination (also a source of genetic variability) is responsible for spreading the mutants throughout the population, and developing new combinations of genetic material with that from 'old' genotypes. A single mutational change may be lost or passed on without great impact on a population, but its effect is modified and enhanced by re-combination.

Indeed, mutation has relatively little effect on variation without the pervasive impact of re-combination.

Genes are relatively stable, though each is known to have a characteristic mutation rate. Genetic mutations are often random occurrences which can involve asexual unicellular organisms, somatic and sexual cells, and the frequency rate of these is solely dependent on the gene. For example in *Neurospora crassa*, the frequency of spontaneous mutation ranges from 3 to 8 per hundred million conidia, and with bacteria, the frequency is in the neighbourhood of $2 \cdot 45 \times 10^{-8}$ bacteria per generation. It is difficult in man to assess the frequency rate. However in the case of haemophilia the haemophilic gene is carried on the X chromosome, and at any given time one-third of the genes are carried by the male population (XY), and the other two-thirds by the female (XX). The disease is manifest in the male and, as a result of this, is subject to extinction, whereas the female is not exposed to the disease since it is recessive and therefore suppressed by its dominant normal allele. Hence with every generation one-third of the haemophilic genes resident in a population risk extinction. Knowing the frequency of the disease within the male population, therefore, it is possible to estimate the number of mutations required to perpetuate the disease. It would seem to be a disease in a steady state — that is, the frequency does not appear to change to any large degree. It has been estimated that a mutation occurs once in every 31 000 individuals in each generation. Thus in this case of a rare sex-linked recessive condition an equilibrium is expected when:

$$\mu = 1/3 \, (1 - f)q$$

where μ is the mutation rate of the normal gene (H) to the haemophilic gene (h) per chromosome per generation; q is the frequency of h in males; and f is the fitness of haemophiliacs, which is the number of haemophiliac progeny compared with the progeny of the rest of the population; the expression 1/3 indicates here that for every haemophiliac gene in males there are two in heterozygous females. The induction of mutation will be discussed at a later stage.

Thus variation due to mutation is seen to occur in what otherwise appears to be a rigidly stable scheme at variable frequencies — since somatic cells of the same and different individuals of the same species are seen to possess identical chromosome numbers and patterns. But just as genes occasionally mutate, so chromosomes exhibit aberrations. The chromosomal aberrations encountered can be classified as follows:

A Variation in Chromosome Number

This group of aberrations can be subdivided further into two major types: (1) those which involve changes in a haploid set (n) as a whole. Euploids are organisms having each of the different chromosomes of the set present in the same number. Examples are monoploid, diploid and triploid which are exact multiples of the haploid chromosome number; (2) those which involve changes (a loss or a gain) in the numbers of chromosomes in a set. Aneuploids are organisms possessing a genetically unbalanced set of chromosomes. That is they have more or less than an integral multiple of the haploid number.

Group 1

Haploidy (n). – Each chromosome is here represented singly. Hence in such a circumstance meiosis is not regular since there is no homologous chromosome present with which to pair.

Polyploidy. – Here a cell possesses more than two haploid sets of chromosomes, since each chromosome is represented by more than two homologues. Instances of polyploidy are quite common in plants, especially with the angiosperms; they are rare, however, in animals.

Triploidy (3n); tetrapolyploidy (4n); and pentapolyploidy (5n). – An auto-polyploid is a polyploid in which all the chromosomes are derived by way of multiplication from the same species (a single diploid); that is, the homologue was homozygous. Allopolyploidy, on the other hand, is a form resulting from a cross between diploid species which possess different chromosome sets with the net result that the hybrid has a chromosome number differing from that of the parental cells. Among others, triploids and tetraploids have been induced experimentally, utilizing temperature shock techniques. Polyploidy, as already stated, is a rarity in animal cells, but can be induced both by the temperature shock method and by substances such as colchicine and acenaphthene which have the ability to prevent spindle formation. As a consequence, following the resumption of normal cell activity after a requisite period of time, the cells possess a double chromosome number.

Group 2

Monosomic. – This is a situation in which one chromosome is lost from one set, and where this concerns a diploid cell the chromosome complement is 2n-1.

Polysomic. – Where there is an addition of one or more chromosomes to one set the aberration is termed polysomic. For example, trisomic which is 2n + 1 (*see* mongolism); tetrasomic which is 2n +2.

Nullisomic. – Here there is a loss in the chromosome complement in one complete pair – that is, 2n-2.

A Variation in Number and Spatial Arrangement of Gene Loci in Chromosome Structure

The types of aberration which may be classified as in this category are as follows:

Translocation. — Here chromosomal rearrangement has occurred. That is portions are exchanged between non-homologous chromosomes resulting in a new chromosome. Furthermore translocation may be both homozygous and heterozygous.

Inversion. — When segments of genes within a chromosome structure are inverted through 180 degrees, inversion is said to have occurred.

Duplication. — This term is used when one or more genes is represented two or more times in a particular chromosome (*Figure 6.8 (3)*).

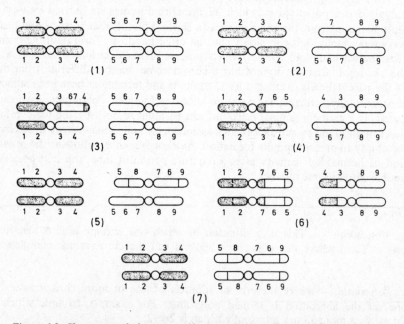

Figure 6.8. Chromosomal aberration: (1) normal; (2) deficiency; (3) duplication; (4) heterozygous translocation; (5) heterozygous inversion; (6) homozygous translocation; (7) homozygous inversion

Deficiency. – This is an aberration in which a terminal or interstitial segment of the chromosome may be lost, and it can involve one or more genes. A terminal deficiency is illustrated in *Figure 6.8 (2)*. For a discussion on the classification and recognition of normal and abnormal karyotypes *see* pages 210 and 217.

Induction of Mutation

Mutations can be induced by either radiation or chemicals, and there is apparently no discernible difference between the type of gene mutation which occurs as a result of induction and those occurring spontaneously. The term radiation embraces x-rays, λ -rays, β -rays, fast and slow neutrons and the ultra-violet light rays. Muller (1927)* discovered the mutagenic effect of radiation (x-rays) using *Drosophila melanogaster,* when he found that mutations could be induced at a rate up to 150 times as fast as their spontaneous occurrence. The most common change occurring in Drosophila is the appearance of white eyes, as distinct from the normal red colour. When inbred these white eye mutations persist. On the basis of the work of Muller and others, it seems likely that spontaneous mutation is precipitated by the same factors as those which induce mutation artificially – namely that all organisms within the confines of the biosphere are subjected to cosmic rays and other forms of radiation, in addition to mutagenic chemicals present within the environment. It has been found that the number of mutations induced by radiation is directly proportional to the dose. The effect of the alkaloid colchicine on the mitotic mechanism has already been mentioned and this has led to further experimentation with other substances, such as phytohaemagglutinin, as a means of studying normal and abnormal mitosis. Such substances acting on the cell nucleus produce chromosomal configurations which facilitate the enumeration and analysis of the karyotype (*see* page 210).

GENES AND CHROMOSOMES

Knowledge of the molecular structure of the chromosomes is still incomplete, particularly with regard to those of the higher plants and animals. Morphologically chromosomes are dense cylindrical structures which can be stained easily with basic dyes. Using the Feulgen stain, DNA can be almost exclusively detected in association with their structure. Their gross morphology is that of a cylinder possessing a constriction which is known as the centromere. According to the position of this centromere so the chromosomes can be classified into three basic types: acrocentric; sub-metacentric; and metacentric (*Figure 6.9*). Other basic characteristics common to the majority of chromosomes is that they divide longitudinally into two halves to form chromatids, and each chromatid consists of one or more threads called chromonemata. The chromonemata are particularly well defined in the lampbrush chromosomes observed in the oocytes of a number of vertebrates and invertebrates.

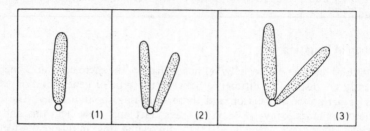

Figure 6.9. The three morphological types of chromosome: (1) acrocentric; (2) submetacentric; (3) metacentric

Generally speaking the fundamental unit of the chromosome structure — the dimensions of which are approximately 100Å — consists of a nucleoprotein assembly; that is, two DNA—histone molecules. The basic protein (histone) appears to be associated with the nucleic acids throughout the entire chromosome length, and it is conjectured by some workers that its intimate association with DNA may play some role in regulating genetic activity. The primary function of the histones, however, is still not clear. Similarly there is evidence that RNA is concentrated within the protein—DNA complex, since it can be demonstrated in the lampbrush chromosomes, and in the giant chromosomes of the Diptera flies (in the salivary glands), where it is in association with structures known as puff or Balbiani rings. These rings are simply expansion areas within the chromonemata. Also associated with the chromosome ultrastructure are the non-histones, the acid proteins; again the significance of these is unknown.

Thus because of the extreme complexity of the chromosomes of higher organisms much indirect information from the study of viral and microbial genetics has been assembled which collectively suggests (though does not necessarily prove) that DNA is the material substance of the genes in higher organisms. In fact the experiments of Meselson and Stahl (1958) using *E.coli* probably offer the most convincing experimental evidence to support the Crick and Watson hypothesis of DNA replication. Basically this work (*Figure 6.10* depicts diagrammatically the Meselson and Stahl experiment), involved growing bacteria (DNA) in the presence of a heavy isotope nitrogen ^{15}N for several generations. The isotope ^{15}N was used as a tracer to follow the distribution of the nitrogenous bases in subsequent generations. The bacteria were then grown on ordinary media containing only ^{14}N. Examining this differential data it was found that the DNA which was initially ^{15}N changed to an intermediate density at the time when most cells were completing their first cell division. That is, a hybrid molecule was formed. Similarly the next generation showed an intermediate and a light band, all of which was expected in successive duplication of the DNA according to the hypothesis of Crick and Watson.

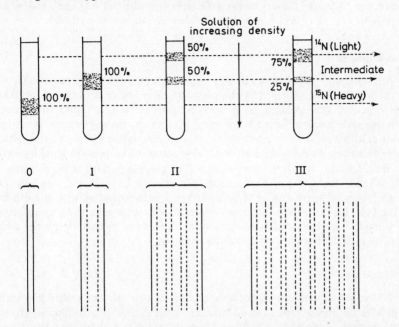

Figure 6.10. Meselson and Stahl's demonstration of the duplication of DNA in Escherichia coli. A culture of E. coli was grown for two days in a medium containing a source of heavy nitrogen (^{15}N). The cells were freed of external ^{15}N by washing and centrifugation; they were then resuspended in a medium with a ^{14}N source. Sampling took place at intervals corresponding to the doubling of the bacterial colony. The DNA in each sample was separated and spun down, using a density gradient technique. By this means it was possible to distinguish heavy, intermediate and light bands of DNA; the proportions of these corresponded precisely to the conception of DNA duplication according to the Watson-Crick pattern (from Coult, 1966; reproduced by courtesy of Longmans, London)

Thus bacteria have been used to facilitate easy genetical analyses, of which probably the most important factor is that workers are able to compress time into minutes rather than days or weeks — for the average life cycle of Drosophila is some 14 days, whereas for an organism such as E.coli it is in the region of 20 minutes. Thus for the remaining part of this section it is proposed to discuss the bacterial chromosome map as a means of appreciating the concept of gene arrangement on the chromosome skeleton.

As seen earlier classical genetics involved work with diploid organisms such as Drosophila and plants in which re-combination is brought about by the formation of a zygote and meiotic division. However in the lower protista re-combination can occur through three processes termed: transformation, conjugation and transduction — all of which reveal a primitive mechanism of sexuality.

Transformation

An artificial process (which may occur naturally on rare occasions) transformation was first observed in the organism *Streptococcus pneumoniae*. When this organism was isolated from infected patients it was found to possess a capsule, and *in vitro* formed smooth (S) colonies. On subsequent serial subculture, however, it manifested rough (R) colonies within the population which, apart from not having capsules, were also found to be non-virulent. By virtue of this capsule, which is composed of a polysaccharide, it is possible to differentiate the types of pneumococci, designated I, II, III, etc. For example, it was found that by mixing R cells of type I pneumococcus with the cell-free extract of S cells of type II, some of the R cells were genetically transformed to type II — revealing that re-combination is effected in bacteria by the direct transfer of DNA from the donor to the recipient.

Conjugation

In the subsequent search for a more definite process, one resembling sexual reproduction in bacteria was undertaken using *E.coli*. It was found that two mating types designated F^+ (males) and F^- (females) exist, and that during conjugation F^+ acts as a genetic donor to the recipient F^-, the mechanism of which requires direct cell contact.

Thus sexuality in the bacterium is determined by the presence of a specific genetic factor called the F agent, which can occur either as a part of the *E.coli* chromosome or as a free chromosome. Those F^+ cells containing the sex factor integrated within the chromosome are called Hfr, or high frequency of re-combination cells, which indicates an increase in transfer efficiency by a factor of 10^3, as compared to the low transfer rate of the other F cells. Hence fundamentally the mechanism of genetic transfer in conjugation involves the F agent becoming attached to the male chromosome forming a transfer-competent cell, now called an Hfr cell. Contact between the Hfr and the F^- cells is achieved by the formation of a cytoplasmic bridge or pili between the two cells. With this, the chromosome of the Hfr cells snaps and duplicates at the site at which the F factor has been integrated. The donor cell then injects a chromosome thread into the F^- cell — a procedure taking approximately 2 hours. With separation, re-combination occurs between the donor and F^- chromosome, and excess genes are eliminated, thereby establishing the haploid condition.

Transduction

The phenomenon of transduction is similar to that of conjugation except that in this case the genetical material is conveyed by means of a bacteriophage. A typical bacteriophage such as the T4 is of complex structure, but basically it consists of a head which contains DNA and a narrow cylinder constituting the tail (*Figure 6.11a, diagram 1*). The tail possesses an outer contractile sheath connected to fibres which initially anchor the phage superstructure to the host

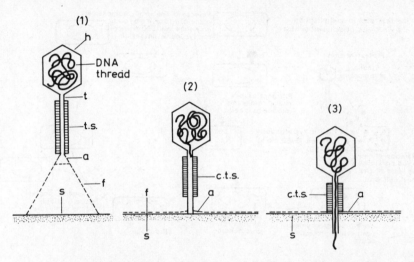

t = tail; t.s. = tail sheath; c.t.s. = contracting tail sheath; h = head; a = anchor; s = bacterial surface; f = attachment fibre

Figure 6.11. (a) Shows probable stages in the infection of a colon bacillus by a T4 phage particle. (1) The phage particle consists of a hollow hexagonal head (h) and a tubular tail (t). Both are known to be proteinous. Within the head lies a much-coiled thread of DNA. Around the tube of the tail is an outer zone or tail sheath of contractile protein (t.s.). The tube ends in attachment fibres (f) which initially attach the particle to the bacterial cell. (2) and (3) The anchor of prongs (a) grips the cell surface, and as the outer protein tail sheath contracts, the tube is driven down across the bacterial wall, and the DNA thread is injected into the cell (from Coult, 1966; reproduced by courtesy of Longmans, London)

cell. The overall structure is rather analogous to a hypodermic syringe and needle, for when contact is made with a bacterium, the attachment fibres secrete an enzyme which partially digests the reception site. The outer sheath then contracts plunging the phage tail cylinder into the cell wall of the host, and then the phage DNA is injected into the bacterium cytoplasm (*Figure 6.11a, diagrams 2 and 3*). From this point on the host metabolism is

193

switched, and committed to the exclusive synthesis of the phage DNA and protein. The result of this is the synthesis of new phage particles; and on achieving maximum replication, the committed cell lyses, releasing the phage progeny. This is the terminal stage in the life cycle of a lysogenic phage. (*Figure 6.11b*). Where the DNA of a phage enters a host cell and does not enter a lysogenic cycle but rather becomes integrated into the host chromosome — that is, becomes latent — the phage is said to be a prophage.

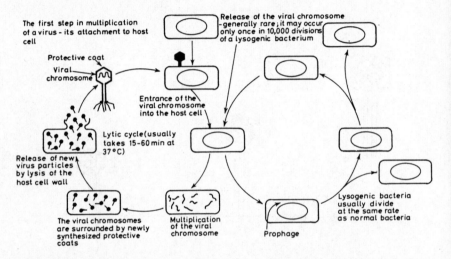

The first step in multiplication of a virus - its attachment to host cell

Protective coat

Viral chromosome

Entrance of the viral chromosome into the host cell

Lytic cycle(usually takes 15-60 min at 37°C)

Release of new virus particles by lysis of the host cell wall

The viral chromosomes are surrounded by newly synthesized protective coats

Multiplication of the viral chromosome

Release of the viral chromosome - generally rare; it may occur only once in 10,000 divisions of a lysogenic bacterium

Prophage

Lysogenic bacteria usually divide at the same rate as normal bacteria

Figure 6.11. (b) The life cycle of a lysogenic bacterial virus. We see that, after its chromosome enters a host cell, it sometimes immediately multiplies like a lytic virus and at other times becomes transformed into prophage. The lytic phase of its life cycle is identical to the complete life cycle of a lytic (non-lysogenic) virus. Lytic bacterial viruses are so called because their multiplication results in the rupture (lysis) of the bacteria (from Watson, 1965 in Molecular Biology of the Gene; reproduced by courtesy of Benjamin, Inc., New York)

Chromosome Map

Using Hfr mutants it is possible to assemble a picture of the spatial order of genes along the bacterial chromosome. The rationale of this is that two genetic markers will remain together during re-combination, and will only become separated by crossing-over; if they are close together, however, the chance of this occurring is remote. It has been found that the F$^+$ strain of *E.coli* has only one linkage group or chromosome, and that it forms a closed circle of genetic markers. Furthermore in each Hfr strain the position of the break in the chromosome circle will vary prior to its insertion into the female receptor cell, since the male chromosome must break in order that a free end can enter the female.

194

If this breakage point were not constant for a given strain, the transfer of a particular gene set prior to others would not be seen. The point of breakage is always specific and appears to occur at the point of F factor integration. For example, in one particular strain of *E. coli* soon after mating, those genes controlling synthesis of leucine and threonine are transferred. This consistency in the breakage point and specificity of linkage along the closed circuit of the chromosome has enabled workers to assemble a chromosome map of *E.coli*. Such a genetic map of *E.coli* using this experimental technique, is assembled in *Figure 6.12*, and a key to the gene abbreviations is given in

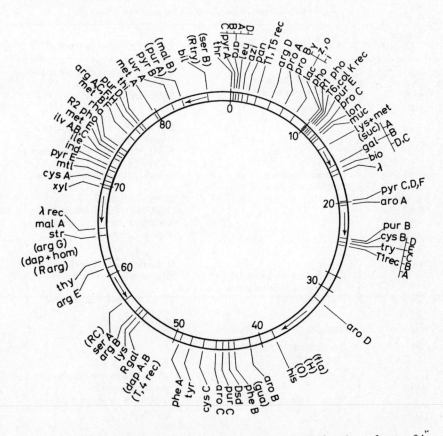

Figure 6.12. The genetic map of E. coli. *The symbols mark the locations of genes. *A key to the various gene abbreviations is given in Table 6.3. (From Taylor and Thoman, 1964; reproduced by courtesy of Genetics, N.Y.)*

Table 6.3. Those genes the location of which are only approximately known are shown in parentheses. The numbers divide the map into time intervals corresponding to the time in minutes which it takes each male chromosomal segment to move into a female cell. Thus 89 minutes are now thought to be required for complete transfer. The arrows mark the points at which various Hfr chromosomes break prior to transfer into a female cell; the direction of the arrows indicates transfer direction.

TABLE 6.3

*Key to the Genes of E.coli Chromosome

Genetic symbols	Mutant character	Enzyme or reaction affected
araD	Cannot use the sugar arabinose as a carbon source	L-Ribulose-5-phosphate-4-epimerase
araA		L-Arabinose isomerase
araB		L-Ribulokinase
araC		
argB		N-Acetylglutamate synthetase
argC		N-Acetyl-γ-glutamokinase
argH		N-Acetylglutamic-γ-semi-aldehyde dehydrogenase
argG	Requires the amino acid arginine for growth	Acetylornithine-d-transaminase
argA		Acetylornithinase
argD		Ornithine transcarbamylase
argE		Argininosuccinic acid synthetase
argF		Argininosuccinase
aroA, B, C	Requires several aromatic amino acids and vitamins for growth	Shikimic acid to 3-enol-pyruvyl-shikimate-5-phosphate
aroD		Biosynthesis of shikimic acid
azi	Resistant to sodium azide	
bio	Requires the vitamin biotin for growth	
cysA	Requires the amino acid cysteine for growth	3-phosphoadenosine-5-phosphosulphate to sulphide
cysB		
cysC		Sulphate to sulphide; 4 known enzymes
dapA	Requires the cell wall component diaminopimelic acid	Dihydrodipicolinic acid synthetase
dapB		N-Succinyl-diaminopimelic acid deacylase

196

TABLE 6.3 *cont.*

Genetic symbols	Mutant character	Enzyme or reaction affected
dap + hom	Requires the amino acid precursor homoserine and the cell-wall component diaminopimelic acid for growth	Aspartic semialdehyde dehydro-genase
Dsd	Cannot use the amino acid D-serine as a nitrogen source	D-Serine deaminase
fla	Flagella are absent	
galA	Cannot use the sugar galactose as a carbon source	Galactokinase
galB		Galactose-1-phosphate uridyl transferase
galD	Constitutive synthesis of galactose operon proteins	Uridine-diphosphogalactose-4-epimerase
galC		Defective operator
gua	Requires the purine guanine for growth	
H	The H antigen is present	
his	Requires the amino acid histidine for growth	10 known enzymes †
ile	Requires the amino acid isoleucine for growth	Threonine deaminase
ilvA	Requires the amino acids isoleucine and valine for growth	α-Hydroxy-β-keto acid rectoisome-rase
ilvB		α, β-dihydroxyisovaleric dehydrase†
ilvC		Transaminase B
ind (indole)	Cannot grow on tryptophane as a carbon source	Tryptophanase
λ	Chromosomal location where prophage λ is normally inserted	
lac Y	Unable to concentrate β-galactosides	Galactoside permease
lac Z	Cannot use the sugar lactose as a carbon source	β-Galactosidase
lac O	Constitutive synthesis of lactose operon proteins	Defective operator

(continued)

TABLE 6.3 cont.

Genetic symbols	Mutant character	Enzyme or reaction affected
leu	Requires the amino acid leucine for growth	3 known enzymes†
lon (long form)	Filament formation and radiation sensitivity are affected	
lys	Requires the amino acid lysine for growth	Diaminopimelic acid decarboxylase
lys + met	Requires the amino acids lysine and methionine for growth	
λ rec, malA	Resistant to phage λ and cannot use the sugar maltose	Phage λ receptor, and maltose permease
malB	Cannot use the sugar maltose as a carbon source	Amylomaltase (?)
metA	Requires the amino acid methionine for growth	Synthesis of succinic ester of homoserine
metB	Requires either the amino acid methionine or cobalamine for growth	Succinic ester of homoserine + cysteine to cystathionine
metF		5,10-Methylene tetrahydrofolate reductase
metE		
mtl	Cannot use the sugar mannitol as a carbon source	Mannitol dehydrogenase (?)
muc	Forms mucoid colonies	Regulation of capsular polysaccharide synthesis
O	The O antigen is present	
pan	Requires the vitamin pantothenic acid for growth	
phe A, B	Requires the amino acid phenylalanine for growth	
pho	Cannot use phosphate esters	Alkaline phosphatase
pil	Has filaments (pili) attached to the cell wall	
proA proB proC	Requires the amino acid proline for growth	

TABLE 6.3 *cont.*

Genetic symbols	Mutant character	Enzyme or reaction affected
purA	Requires certain purines for growth	Adenylosuccinate synthetase
purB		Adenylosuccinase
purC, E		5-Aminoimidazole ribotide (AIR) to 5-amino-imidazole-4-(N-succino-carboximide) ribotide
purD		Biosynthesis of AIR
pyrA	Requires the pyrimidine uracil and the amino acid arginine for growth	Carbamate kinase
pyrB	Requires the pyrimidine uracil for growth	Aspartate transcarbamylase
pyrC		Dihydroorotase
pyrD		Dihydroorotic acid dehydrogenase
pyrE		Orotidylic acid pyrophosphorylase
pyrF		Orotidylic acid decarboxylase
R arg	Constitutive synthesis of arginine	Repressor for enzymes involved in arginine synthesis
R gal	Constitutive production of galactose	Repressor for enzymes involved in galactose production
RI pho, R2 pho	Constitutive synthesis of phosphatase	Alkaline phosphatase repressor
R try	Constitutive synthesis of tryptophane	Repressor for enzymes involved in tryptophane synthesis
RC (RNA control)	Uncontrolled synthesis of RNA	
rha	Cannot use the sugar rhamnose as a carbon source	
serA	Requires the amino aicd serine for growth	3-Phosphoglycerate dehydrogenase
serB		Phosphoserine phosphatase
str	Resistant to or dependent on streptomycin	
suc	Requires succinic acid	
T1, T5 rec	Resistant to phages T1 and T5 (mutants called B/1, 5)	T1, T5 receptor sites absent
T1 rec	Resistant to phage T1 (mutants called B/1)	T1 receptor site absent

(continued)

TABLE 6.3 cont.

Genetic symbols	Mutant character	Enzyme or reaction affected
T6, colk rec	Resistant to phage T6 and colicine K	T6 and colicine receptor sites absent
T4 rec	Resistant to phage T4 (mutants called B/4)	T4 receptor site absent
thi	Requires the vitamin thiamin for growth	
thr	Requires the amino acid threonine for growth	
thy	Requires the pyrimidine thymine for growth	Thymidylate synthetase
tryA		Tryptophane synthetase, A protein
tryB		Tryptophane synthetase, B protein
tryC	Requires the amino acid tryptophane for growth	Indole-3-glycerolphosphate synthetase
tryE		Anthranilic acid to anthranilic-deoxyribulotide
tryD		3-Enolpyruvylshikimate-5-phosphate to anthranilic acid
tyr	Requires the amino acid tyrosine for growth	
uvrA	Resistant to ultra-violet radiation	Ultra-violet-induced lesions in DNA are reactivated
xyl	Cannot use the sugar xylose as a carbon source	

*Each known gene or gene cluster is listed by its symbol and with the character caused by a mutation in the gene or gene cluster. The enzyme affected or reaction prevented is listed where known.

†Denotes enzymes controlled by the homologous gene loci of Salmonella typhimurium. (Modified from Taylor and Thoman, 1964; reroduced by courtesy of Genetics, N.Y.)

THE GENETIC CODE

The salient feature of DNA is its double right-handed helical structure consisting of two polymeric chains, the composite diameter of which is approximately 20 Å, each chain making a complete revolution at every 34 Å..Each chain is a polynucleotide in which the resident sugar moiety is linked by a phosphate group to a sugar of an adjacent nucleotide. It has been found that there are 10 nucleotides per chain in every turn of the helix and, as a consequence, the distance per nucleotide base is 3·4 Å. Each nucleoside is arranged in such a manner that it is perpendicular to the nucleotide chain, and the two chains are held by hydrogen bonds established between the base pairs. Pairing of the bases is highly specific — one purine with one pyrimidine. There are two pyrimidines — thymine (T) and cytosine (C); and two purines — adenine (A) and guanine (G). Thus A-T, C-G, T-A and G-C are the only possible pairing arrangements. This strictness of the pairing results in a complementary relation between the sequence of bases in the two chains — that is if a sequence on one chain is A,T,G, T, and C, then the opposite must be T,A,C,A and G.

The replication mechanism of DNA as postulated by Watson and Crick (1953) involves separation of the helices by unwinding, and the formation of complementary molecules on each of the free single strands by a template mechanism. Thus it is that the structure of DNA in relation to this replication requires the unwinding of the helix at a certain point, and the presence of an enzyme DNA polymerase which dictates nucleotide linkage in polynucleotide production. This mechanism suggests that during replication half of the DNA helix is conserved: this hypothesis is supported by the experimental results of Meselson and Stahl (1958) and Taylor, Woods and Hughes (1957).

The DNA molecule (the chromosome) of a bacterium (for example, E.coli) has been found to be circular. The hypothetical reason for this is the prevention of surplus replication, for during cell division the chromosome must only replicate once. This breakage of the chromosome circle occurs only during the division cycle with replication occurring at one of the terminal break points. The mechanism controlling such a system, is completely unknown.

In the higher organisms genes are arranged in a sequential order along the chromosome superstructure. The central question is, whether the gene can be considered as a distinct pair of DNA molecules, or as a portion of a larger unit. Similarly whether there is only one copy of each gene per chromosome, or more than one. Though precise knowledge of the chromosome structure is lacking, the most favourable chromosome model depicts the genes as pairs, which are aligned side by side, perpendicular to the chromosome axis, with their ends joined together by protein molecules.

The work of Taylor, Woods and Hughes (1957) and Taylor (1959b) assembled further support for the Watson and Crick replication hypothesis. Fundamentally the experiments were devised to demonstrate the distribution of new DNA between two daughter chromosomes and between the chromosomes formed from each of them following further division — using as models the root tips of plants. Initially these root cells were immersed in a solution of nucleotide thymidine labelled with radioactive hydrogen ^{3}H. After treatment with ^{3}H-thymidine, and using a special autoradiographic technique, Taylor found that each pair of daughter chromosomes contained equal amounts of radio-active DNA. Hence during duplication of a chromosome, it yields two pairs, containing one old and one new molecule each — here one inactive and one radioactive. Thus experimentally replication occurred as predicted. DNA acts as a template for its own replication — each strand giving rise to a complementary ribbon with the formation of two doubles, after the completion of one synthetic cycle.

The concept of a linear relationship between the nucleic acids and protein was first postulated by Pontecorvo (1952), and from subsequent studies of DNA structure culminating in the publication by Watson and Crick in 1953 of their paper, 'Molecular Structure of Nucleic Acids' and the differential structure studies of normal and abnormal proteins, concrete evidence evolved supporting the theory that the amino-acid sequence in a protein chain was initially determined by the nucleotides resident in the DNA molecule. Fundamentally the Crick and Watson hypothesis was that genetic information of inheritance was encoded in the base sequence present in each of the 46 chromosomes and that each triplet sequence represented a message for the assembly of a specific amino acid. As a result of this, several code theories arose which by analogy equated the four bases resident in the DNA molecule with a minimum dictionary containing 20 words, since some 20 amino acids occur in those protein structures so far elucidated. Obviously, one letter would only be sufficient to take care of four of the 20 amino acids: that is, the letters A, T, C or G give only four possibilities. Similarly the second letter of the word could be A, T, C, or G, which makes possible (4 X 4) or 16 different words: AT, TC, TG, GT.... It can be seen immediately that this is still an inadequate number. If, however, there are three bases coding for each amino acid, the number of possible triplets is 64 (that is 4 X 4 X 4), which is more than enough to code each amino acid with a surplus.

The next problem under consideration concerned the nature of the reading mechanism, and exactly how the end of one word was distinguished from the beginning of the next — since there must be some restriction on which words constitute sense and which nonsense. For instance, taking an apparently undifferentiated sentence like: ATTCGCATAGCA.... this code could quite easily be read either as: A, TTC, GCA, TAG.... or as ATT, CGC, ATA.... There are in fact two possible general types of linear triplet codes, an overlapping and a non-overlapping one. In the overlapping triplet codes the first, second and third symbols of the nucleotide sequence code for one message and the second, third and fourth for the second message and so on.

$$A\ T\ T\ C\ G........$$
$$1\qquad 2$$

Alternatively in a non-overlapping code the first, second and third symbol will code for the first message, and the second message, instead of overlapping, is coded by the fourth, fifth and sixth symbols. It would seem according to data at present available that a non-overlapping code is the code which has arisen in nature. For if a mutation occurred changing one letter in a base sequence of an overlapping type, the expectation is that changes in more than one amino acid would result; in fact, experimentally, this has been shown not to be the case. Furthermore such a scheme as the overlapping type does not permit a certain frequency of amino acids occurring together.

Using acridine orange to induce mutations in the B1 segment of the B cistron of the rII locus of the T4 phage Crick *et al.* (1961) found that the beginning of each genetic sentence was determined by a marker, and that the words were triplets. The experiment can be summarized as follows: assuming that a normal base sequence reading from left to right was CAT/CAT/CAT/CAT/CAT. . . . if by chance a mutagen added or deleted a base from a nucleotide sequence, then the resultant sequence is seen to be recognizably defective. That is, with an extra base it would read CAT/GCA/TCA/TCA/TCA/T, and where a deletion occurred the effect would be to produce the sequence CAT/CAT/CTC/ATC/AT. But if a base is added and another deleted as a result of re-combination-restoration or the reversion to a message closely resembling the original the sequence would appear as CAT/GCA/TCT/CAT/CAT. The result of further work along these lines seems to favour the concept of a non-overlapping triplet code which requires a marker that will dictate the beginning of each code sentence at the correct place. Thus a particular triplet sequence commences from a fixed point: if somewhere along the ensuing sequence a base is removed, substituted or duplicated, then the code becomes meaningless for the remainder of the sequence.

The major problem now facing molecular biologists is the deciphering of the code and the recognition of a message. Nirenberg and Matthaei (1961) found that adding polyuridylic acid to a ribosome system results in the synthesis of polyphenylalanine — a protein consisting of one amino acid — suggesting that a series of uridylic acid residues in the RNA coded for phenylalanine, and on the strength of a triplet code, the triplet for phenylalanine would spell UUU. Exploiting this technique Ochoa and Nirenberg have independently shown that the code is degenerate; that is, that the individual amino acids would be coded by more than one triplet, and which codon represents which amino acid in terms of the sequence of bases on the mRNA. As Dr. Jukes has pointed out in his chapter in *Mammalian Protein Metabolism* the expression of minimum base differences per codon is based on the genetic code and his table of the code is reproduced here (Table 6.4; *see also* Jukes, 1962; 1963a, b; 1969)

Clearly then the genetic code does not conform rigidly to our present ideas of cryptography. Indeed much speculation has arisen as to the possible function of, or reason for, degeneracy and nonsense of the code. With regard to the question of degeneracy a vestigial pattern appears to emerge, in which the doublet CG is often shared by the amino acids arginine and alanine, but the significance of this is still unknown (Ochoa, 1964). In an attempt to understand

203

this phenomenon Woese (1965) interpreted degeneracy from a functional point of view; namely, that the code had specific degeneracies confined to particular areas of the codon — acting, he tentatively suggested, as a kind of cryptic punctuation. Furthermore he was able to show, as a result of chromatographic

TABLE 6.4

The Genetic Code

UUU Phenylalanine	CUU Leucine	AUU Isoleucine	GUU Valine
UUC Phenylalanine	CUC Leucine	AUC Isoleucine	GUC valine
UUA Leucine	CUA Leucine	AUA Isoleucine	GUA Valine
UUG Leucine	CUG Leucine	AUG Methionine	GUG Valine
UCU Serine	CCU Proline	ACU Threonine	GCU Alanine
UCC Serine	CCC Proline	ACC Threonine	GCC Alanine
UCA Serine	CCA Proline	ACA Threonine	GCA Alanine
UCG Serine	CCG Proline	ACG Threonine	GCG Alanine
UAU Tyrosine	CAU Histidine	AAU Asparagine	GAU Aspartic acid
UAC Tyrosine	CAC Histidine	AAC Asparagine	GAC Aspartic Acid
UAA chain termn.	CAA Glutamine	AAA Lysine	GAA Glutamic acid
UAG chain termn.	CAG Glutamine	AAG Lysine	GAG Glutamic acid
UGU Cysteine	CGU Arginine	AGU Serine	GGU Glycine
UGC Cysteine	CGC Arginine	AGC Serine	GGC Glycine
UGA chain termn.	CGA Arginine	AGA Arginine	GGA Glycine
UGG Tryptophane	CGG Arginine	AGG Arginine	GGG Glycine

(From *Evolution of Protein Molecules,* 1969; reproduced by courtesy of the author Dr T. H. Jukes and Academic Press.)

studies (Rf values), a relationship between the amino acids showing particular codon bases. This work would seem to indicate that degeneracy is a code within a code. In the case of nonsense this enigma — which refers to a base triplet not apparently coding for an amino acid — has been interpreted as a triplet carrying intrinsic punctuation instructions: such, for example, as 'begin here and end there'. But whatever the true solution is they are obviously interrelated. The reader is advised to consult a paper by Platt (1962) who endeavours to draw an intriguing analogy between the transfer of genetic information and a complex instruction manual — thereby establishing a coherent and workable scheme for coming to grips with this baffling phenomenon.

PROTEIN SYNTHESIS

The process of protein synthesis involves fundamentally the conveyance of specific coded information derived from DNA resident within the interphase nucleus, which gives rise to 3 types of RNA – messenger RNA (mRNA), ribosomal RNA (rRNA) and transfer RNA(tRNA). Hence DNA is not a direct template for protein synthesis, since the genetic information of DNA is transferred via intermediaries which act as the primary templates dictating the sequential order of amino acids in the polypeptide chain. Structural similarities between DNA and RNA are discussed in Chapter 3. It will be remembered that the sugar of DNA is deoxyribose, whereas RNA contains ribose; also RNA contains a related pyrimidine called uracil as opposed to thymine present in DNA.

Thus in general the process involves RNA transcription upon the DNA template. Initially amino acids (AA) are activated by enzymes which become linked to tRNA forming a complex AA–tRNA which is then transported to a ribosome site situated in the endoplasmic reticulum. At this site the AA–tRNA complex which carries a triplet of bases or anticodon recognizes its counterpart, and in consequence forms a temporary bond with a complementary sequence, the codon, on the mRNA. The insertion of a specific amino acid is determined simply by the mRNA. Thus a polypeptide chain is slowly assembled as illustrated in *Figure 6.13*.

The precise function of rRNA is still largely unresolved, though it is known to form an integral part of the ribosome structure, which consists of the mysterious rRNA and protein. In the presence of a low magnesium concentration the spherical ribosomes can be split into two fractions of 30S and 50S respectively. During protein synthesis rRNA is also known to be intimately involved in the linking of tRNA and mRNA to the ribosome particle – for at the commencement of synthesis the ribosome becomes attached to the leading end of the mRNA, which as it advances along the mRNA strand, determines the sequence of amino acids as indicated in *Figure 6.13*. Hence, though the assembly role of the ribosome in protein synthesis has been established, the precise mechanism involved is still unknown. Furthermore mRNA cannot act as a template until it is firmly attached to the ribosome; also each mRNA is found in association with a group of ribosomes, which are capable of moving along its length and in some way interpret the resident code.

Messenger RNA is copied from the corresponding locus of the genome, and is transferred to the endoplasmic reticulum where it becomes associated with a number of ribosomes, constituting what is known as a polyribosome. It is believed that either the DNA bases swing out, thus becoming available on the exterior of the spiral form to allow transcription to occur; or that the information is copied from a section of the DNA molecule which has temporarily unwound, thus exposing specific strands. But by virtue of the nature of the mechanism it is clear that only one of the two strands is used as a template for mRNA synthesis.

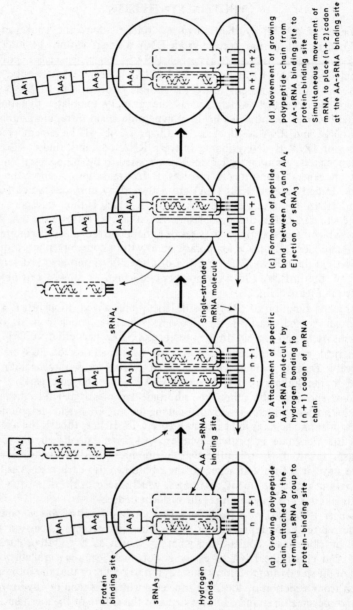

Figure 6.13. Diagrammatic representation of the stepwise growth of a polypeptide chain (from Watson, 1965 in Molecular Biology of the Gene; reproduced by courtesy of Benjamin, Inc., New York)

It is assumed, therefore, that each particular strand of mRNA possesses a base distribution complementary to a specific area of DNA from which it was copied, differing only in so far as the mRNA uracil substitutes thymine. It has been shown that an enzyme RNA polymerase catalyses the synthesis of mRNA in the presence of DNA:

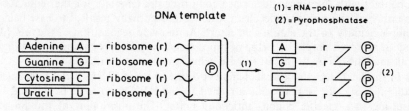

This enzyme unites the ribonucleotides by catalysing the formation of $3'$, $5'$-phosphodiester bonds which hold the RNA structure together. Much of the evidence concerning the nature and formation of mRNA has been gathered from the following fundamental observations: (1) a hybrid RNA:DNA complex can be isolated in which the RNA can be labelled by adding labelled precursors; (2) labelled RNA has base ratios similar to those of the parent DNA; (3) this same material binds reversibly with ribosomes; (4) there is a rapid turnover of RNA in the cell.

A single polynucleotide chain, tRNA contains 70–80 nucleotides, and has a molecular weight in the region of 25 000. Several structures have been proposed (Levitt, 1969), the conformation of which make it specific for a particular type of amino acid. The terminal sequence of the nucleotide to which the specific amino acid becomes attached has been shown to be always cytidylic–cytidylic–adenylic residues (C–C–A). With regard to the synthesis of tRNA, the proposition has been advanced that RNA formed during the time whilst DNA was unwound gave rise to a messenger and the complementary strand became tRNA; as yet there is no substantial experimental evidence to support such a theory.

Amino acids are activated by reacting with ATP in the presence of a specific activating enzyme called amino-acyl-synthetase. In this activation reaction the amino acid is esterified to form a high-energy linkage with a specific amino-acid carrier molecule; the activating enzyme then transfers the amino acid to the terminal adenylic acid residue of the tRNA:

$$\text{Amino acid (AA)} + \text{ATP} \underset{}{\overset{\text{Amino acyl synthetase}}{\rightleftharpoons}} \text{AA} \sim \text{AMP} + \text{(P)}-\text{(P)}$$

$$\text{AA} \sim \text{AMP} + \text{tRNA} \underset{\text{synthetase}}{\overset{\text{Amino acyl}}{\rightleftharpoons}} \text{AA} \sim \text{tRNA} + \text{AMP}$$

Each of the 20 amino acids has a specific activating enzyme and tRNA carrier molecule.

Following activation the AA–tRNA moves to the ribosome on which the peptide bonds form. During this time mRNA has become associated with a ribosome particle complex – five or more. That portion of the mRNA molecule in contact with the ribosome surface comprises the template for the sequential assembly of the polypeptide chain which will eventually result, the base being complementary to the amino-acid insert. Amino acid recognition of the mRNA code is achieved simply by virtue of it possessing specific recognition nucleotides built into the tRNA structure, the site being known as the anticodon.

In the next step, the spatial alignment of the series of tRNA carrier molecules with their specific amino acids attached to the mRNA molecule is required – resulting in a specific codon–anticodon union. The positioning of the amino acids and their subsequent transference from the AA–tRNA complex involves an ATP-like molecule called guanosine triphosphate (GTP). Thus a polypeptide chain grows by stepwise addition of single amino acids, which are assembled on the template starting with the amino terminal end. A peptide bond is formed immediately following the release of tRNA from the carboxyl terminal amino acid. The incomplete molecule remains attached to the template by the preceding amino-acid residue. This proceeds until the polypeptide chain has grown to the desired length, whereupon the terminal tRNA is ejected, a step requiring the participation of ATP – thus creating a free terminal carboxyl group. Thus at one end of each complete polypeptide chain is an amino acid bearing one free carboxyl group and at the other end a free amino group.

CONTROL OF PROTEIN SYNTHESIS

The preceding section reviewed briefly how coded information in the DNA molecule is translated into a specific protein molecule, but no indication was given as to how the whole process was controlled, as clearly it must be controlled. In fact the impression may well have been gained from the text so far that the whole mechanism was simply a stable and unidirectional phenomenon. But this is not the case, for the metabolic activity of a cell which involves such complex events as glycolytic and oxidation cycles, fatty acid and amino-acid metabolism – all of which are closely interrelated – must, in order to maintain the *status quo* with its environment, exhibit a rigid regulatory mechanism, since in normality the cellular components of an organism are seen to remain constant. Indeed this stability of composition is a function of the relative rates of the net intra-cellular reactions. If this were not so then metabolic imbalance would soon result: that is, the rate of formation of one particular intermediary metabolite would exceed that of others. Thus the cell manifests homoeostatic functions in which the cell system is able to adapt, by means of subtle regulatory and co-ordinated metabolism, to changing intra- and extra-environmental conditions, thus maintaining the system in a dynamic steady state. A comprehensive review of this subject is undertaken in Chapter 8.

Jacob and Monod (1961) proposed a scheme which would explain the regulation of protein synthesis; this suggested that the rate of formation of a protein was directly related to the rate of formation of mRNA, and that this is the primary rate-controlling factor. It is thought that structural proteins give rise to mRNA which initiates the synthesis of specific proteins, since the structural gene is that portion of the DNA molecule bearing the code determining the eventual constituent of the protein to be synthesized. The structural gene, it is postulated, is controlled by a master gene called the operator gene, which in order to function requires close linkage, and as a configuration, with a structural gene. A complex of an operator gene and a structural gene is called an operon. The action of the operator gene is controlled by a regulator gene which produces a substance called a repressor, this acting directly on the operator substance by preventing its normal function. Conversely this repressor substance can be inactivated by an inducer substance, so that the repression of the operator is removed, thus permitting the structural gene to be 'switched on'. Hence, the whole scheme of induction and repression rests on the ability of the regulator gene to respond to concentrations of substrates or to products of enzymes produced by the specific operon and under its jurisdiction. Thus an elevation in the concentration of a substrate motivates the regulator gene to induce the production of an enzyme by the removal of repression of the operator. Similarly as the products of the enzymes rise in concentration the process is reversed by repression of the structural gene via the operator gene.

Experimental evidence of the Jacob and Monod hypothesis has been obtained from the predictive studies of mutations affecting different genes in the inducible lactose system of *E.coli* (Monod, Changeux and Jacob, 1963). Designating an active gene as enz$^+$, and its mutant allele as enz$^-$, the operator gene as op$^+$ and its non-functional allele as op$^-$ it can be shown that two heterozygous diploid conditions are possible:

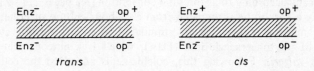

As stated earlier only the *cis* configuration permits enzyme function. In the β-galactosidase-permase system in *E.coli*, there are two genes linked and regulated by the same operator. The experiments of Monod and his colleagues revolved round the mutation effect on the regulator and operator segment in the *E.coli* β-galactosidase-permase operon. The results of these experiments confirm that the operator acts directly on neighbouring structural genes on the same chromosome (requiring both linkages and *cis*-configuration, it should be noted). The Jacob and Monod hypothesis is summarized diagrammatically in *Figure 6.14*. The reader's attention is also drawn to Chapter 2.

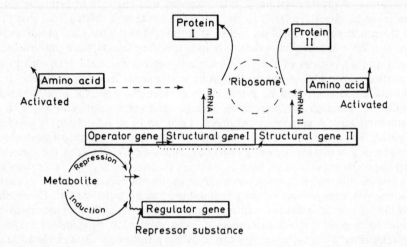

Figure 6.14. Diagrammatic representation of mechanism of regulation of protein synthesis as postulated by Jacob and Monod. This theory suggests that the operator gene activates each structural gene, I and II: The complex is called an operon (see text for full explanation)

CLASSIFICATION AND RECOGNITION OF NORMAL CHROMOSOMES

To recapitulate, a convenient morphological classification of the chromosomes is that utilizing the position of the centromere and, on this basis, it is possible to recognize three types: metacentric, sub-metacentric and acrocentric. A karyotype is the full chromosome complement of an individual cell arranged systematically according to a structural convention (*see* pages 211–214).

Initially a karyotype is prepared from a 72-hour culture of either peripheral blood or marrow cells, which are stimulated to enter a mitotic cycle by the addition of phytohaemagglutinin (PHA), which is a mucoprotein extract of *Phaseolus vulgaris*. Following this, colchicine is added to the cell culture to increase the yield of mitotic cells in the metaphase by preventing the formation of the spindle. The net result is that the chromosomes remain located on the equatorial plane. The culture is then exposed to a hypotonic environment resulting in separation of the chromosomes from the cell structure. They are then fixed and dispersed on to a glass slide by either heat or some other technique. The preparation is then stained for examination and counting. Usual practice is to photograph a suitable microscopic field which can then be cut out and and arranged in homologous pairs and mounted (*Figure 6.15*).

It will be seen that karyotyping requires extreme patience and, most important, good preparations which do not exhibit chromosome overlap, or excessive chromatid separation, or such distortions as over-contraction by colchicine. Clearly then, time spent in perfecting this technique, and in searching for a microscopic field showing good arrested metaphase will yield rewarding results (Tjio and Levan, 1956; Ford and Hamerton, 1960; Nowell, 1960).*

In man the somatic cell possesses a diploid (2n) chromosome number of 46 chromosomes, consisting of a pair of sex chromosomes and 22 other pairs known as the autosomal chromosomes. By convention the 22 autosomal pairs are arranged in order of decreasing size, and then subdivided into 7 groups designated A, B, C, D, E, F, and G (Conference, 1960, 1963).

Normal Human Karyotype

Group A (1–3). – These are the large chromosomes with approximately metacentric centromeres, and the group consists of three pairs.

Group B (4–5). – This group is also made up of large chromosomes, which are characterized by submetacentric centromeres. The two pairs comprising this small group are difficult to distinguish; generally, it is said, 4 > 5 in size.

Group C (6–12). – These are medium sized chromosomes possessing submetacentric centromeres. As with all the autosomes, difficulty is encountered in identifying individual chromosomes. Identification can be aided by using autoradiography which utilizes radioactive substances which can be incorporated into the new molecules of DNA during interphase replication. By virtue of differences in the timing of DNA synthesis in the chromosome species it is possible to differentiate them. Thus 4 completes DNA synthesis much later than does 5 (Yunis, 1965). Also as seen in *Figure 6.15a*, there is included in this group an additional chromosome which is the X chromosome – whereas in *Figure 6.15b*, this Group C contains XX chromosomes.

Group D (13–15). – This group are also medium sized chromosomes with a near-terminal centromere, or acrocentric chromosomes. Members are often referred to as the large acrocentrics, as compared to those of Group C. Once again extreme difficulty is encountered in determining the differences between the three pairs of this group, though it would appear that the Drets and Shaw technique may resolve these problems.

*Two cytologists at Texas University have devised a technique enabling clearer identification of individual chromosomes (Drets and Shaw, 1971). Staining with Giemsa after alkaline treatment produces specific banding facilitating compilation of a karyotype.

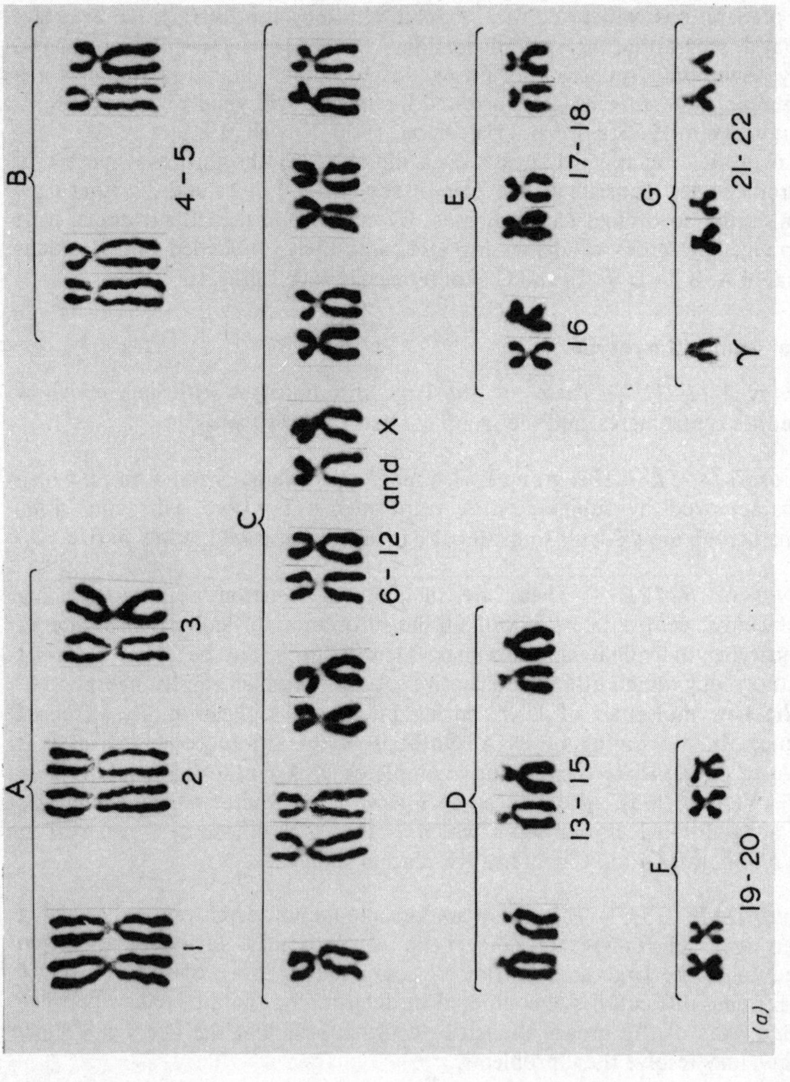

Figure 6.15. (a) Chromosomes of a normal male somatic cell. Note dissimilar sex chromosomes – X and Y – the former not readily distinguishable from the rest of the C group

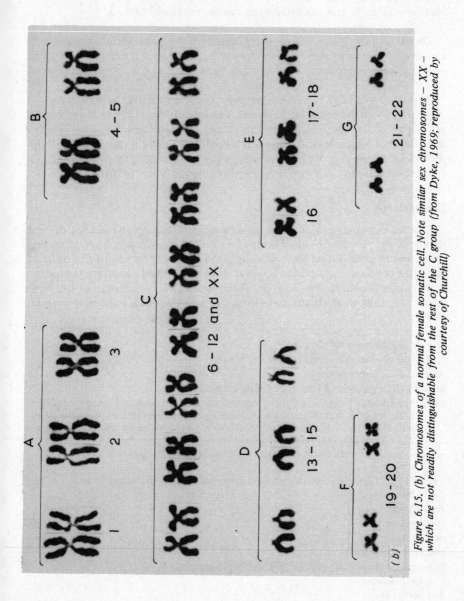

Figure 6.15. (b) Chromosomes of a normal female somatic cell. Note similar sex chromosomes – XX – which are not readily distinguishable from the rest of the C group (from Dyke, 1969; reproduced by courtesy of Churchill)

Group E (16–18). – These are short chromosomes possessing metacentric centromeres in 16, and submetacentric centromeres in 17 and 18.

Group F (19–20). – Two short chromosome pairs with approximately metacentric centromeres distinguish this group, and again it is virtually impossible on morphological grounds to differentiate the two pairs.

Group G (21–22). – These are very short acrocentric chromosomes. In *Figure 6.15* can be seen the Y chromosome which more or less resembles the autosomes of Group G, but often can be distinguished from them by virtue of size, for it is longer than 21 and 22. Recently however a technique has been divised for identifying Y chromosomes in the human interphase nuclei, using a fluorescent acridine derivative (Pearson, Bobrow and Vosa, 1970).

Nomenclature

Due to the rapid proliferation in knowledge of cytogenetics during the last decade it was proposed by the Chicago Conference on Standardisation in Human Cytogenetics (1960), that the following symbols should be adopted in order to minimize confusion in reports. The system thus proposed, describes the human chromosome complement and its abnormalities. '. . . .utilising designations which can easily be understood, written and coded for data retrieval' (Bergsma, 1966).

A – G: the chromosome groups

1–22: the autosomal numbers (the Denver system)

X, Y: the sex chromosomes

diagonal (/): separates cell lines in describing mosaicism

plus (+) or minus (–) sign: when placed immediately after the autosome number or group letter indicates that the particular chromosome is extra (or missing). When placed immediately after the arm or structural designation indicates that the particular arm or structure is larger (+) or smaller (–) than normal

question mark (?): indicates questionable identification of chromosome or chromosome structure

asterisk (*): designates a chromosome or chromosome structure explained in text or footnote

ace: acentric

cen: centromere

dic: dicentric

end: endoduplication

h: secondary constriction or negatively staining region

i: isochromosome

inv: inversion

inv $(p^+ q^-)$
or   pericentric inversion
inv $(p^- q^+)$
mar: marker chromosome
mat: maternal origin
p: short arm of chromosome
pat: paternal origin
q: long arm of chromosome
r: ring chromosome
s: satellite
t: translocation
tri: tricentric
repeated symbols: duplication of chromosome structure

SEX AND AUTOSOMAL CHROMOSOMES

Females are known to possess two X chromosomes, whereas the male possesses one X and one Y chromosome. The male phenotype is thus determined by the presence of the Y chromosome, which apparently carries genes that promote testicular development *in utero*. In the absence of this Y chromosome the foetal testis does not develop; a female phenotype develops in result. Where two or more X chromosomes are present, one is active while the other remains dormant — that is, genetically inert. This statement forms the basis of the Lyon hypothesis. It further states that the X chromosome which is to remain active is determined by chance shortly after conception — an estimated 12–20 days. Hence the active X chromosome may be either matroclinous or patroclinous (that is, maternally or paternally derived). Since the female is mosaic, some cells will have in almost equal numbers either patroclinous active X chromosomes or matroclinous ones. Generally this apparent disparity in the cell population is difficult to detect, though there are exceptions. For instance, where the gene for the erythrocytic enzyme glucose-6-phosphate dehydrogenase is carried on one X chromosome and not on the other, it is possible to detect two distinct erythrocytic populations.

In 1949, Barr and Bertram, whilst studying nerve cells of cats, observed a difference between the nuclei of the male and female cells — namely, that the female cell possessed a small chromatin body attached to the interior wall of the nuclear membrane, which was absent in the male. This sex chromatin or Barr body as it is termed, is apparent in the interphase nucleus. Thus, in cells having more than one X chromosome, this small chromatin mass, which is approximately 1μ in diameter, can be observed lying against the interior surface of the nuclear wall — and where there is only one chromosome this will not be observed. Studies have revealed that the number of sex chromatin corpuscles present at interphase is equal to nX-1; that is, there is one Barr body less than the number of X chromosomes. But in practice this observed relationship is by no means infallible. The X chromosome forming the sex chromatin is considered to be genetically inactive, and this has been confirmed in mice (Ohno, 1961); the results of this work strongly support the Lyon hypothesis. That X chromosome which is visible is termed heteropyknotic, whilst the active X chromosome is called isopyknotic.

In 1954, Davidson and Robertson-Smith found that approximately 3 per cent of the female polymorphonuclear neutrophils exhibited a nuclear expansion projecting from the main mass of nuclear lobes and morphologically resembling a drumstick. This is not demonstrable in males. As with Barr bodies, these nuclear extensions can be used to indicate the number of X chromosomes — though, in the case of a situation such as XXX or XXXX one does not necessarily observe two or more drumsticks and, indeed, if after observing 500 cells one does not do so it can be safely concluded that only one X chromosome is present, and if they are detected at a frequency of, say, 3 per cent, then one may conclude that the cells will have at least two X chromosomes present.

The X chromosomes are of the submetacentric type which fall, it will be remembered, into Group C, whereas the Y chromosomes are acrocentric and are to be found in Group G. With regard to the Y chromosome, like the X chromosome it is late in replicating and can be identified using an autoradiographic technique or fluorescent acridine dye as described by Pearson, Bobrow and Vosa (1970). (*See also* footnote, p. 211.) Not a great deal is known about the Y chromosome, other than that it carries the genes determining masculinity.

Sex Chromosome Anomalies

Aberrations of the sex chromosomes are the product of: (1) structural changes, as in the case of translocation; (2) changes arising during meiosis, as for example non-disjunction; or (3) the result of mutagenic agents such as LSD or radiation.

Gonadal Dysgenesis (Turner's Syndrome)

In this syndrome the person (apparently a female) exhibits the following characteristics: short stature; absence of secondary sex characteristics following puberty; webbing of neck; cubitus valgus (an increased carrying angle of elbows); usually vestigial ovaries — such that in later life there is a primary amenorrhoea. The incidence of Turner's syndrome is 1 per 2 500 live births. The karyotype shows a count of 45 chromosomes of the XO pattern (44 autosomal plus X). In 60 per cent of cases no sex chromatin is demonstrable The following other mosaics manifesting Turner's syndrome have been found: XO/XX; XO/XX/XXX; XO/XXX/XO/XY; XO/XYY.

216

Klinefelter's Syndrome

This is a syndrome affecting apparent males who are individuals of a typically asthenic appearance. They are tall with a disproportionate length of limb and the external genitalia are smaller than is normal. Facial hair is slow to develop and pubic hair is sparse while puberty is often delayed. Generally the individual with Klinefelter's syndrome is mentally retarded and often exhibits behavioural difficulties, and social maladjustment. The karyotype shows a count of 47 chromosomes of the XXY pattern (44 autosomal, and XXY). A majority of cases have a positive sex chromatin pattern. The following mosaic patterns have been observed in Klinefelter's syndrome: XXY/XX; XXY/XY; XXXY/XXXXY; and XXXY/XXXXY/XXXXXY.

Autosomal Anomalies

The autosomal anomalies are less frequent than those involving the sex chromosomes, and in the main this is probably due largely to the fact that autosomal defects are predominantly incompatible with normal somatic development.

Down's Syndrome (Mongolism)

This syndrome is commonly associated with an extra chromosome in the G group. The chromosome count in this situation is thus 47. In some cases, however, the count is 46 — the additional chromosome material being translocated into the short limbs of another chromosome group, either group G or D. In this case of familial translocation mongolism, one parent, usually the mother, is found to have a chromosome count of 45. Apparently she only has one No. 21 chromosome, and will be phenotypically normal.

Cri-du-Chat Syndrome

Here in the cri-du-chat syndrome, the karyotype exhibits a count of 46 with an XY pattern, but shows a deletion of the short arm of one of the B group chromosomes (No.5). Characteristics of this syndrome, sometimes known as Lejeune's syndrome, are: microcephaly; oblique palpebral fissues, low-set ears, beaked nose and micrognathia. Also these patients have a distinctive high pitched cry which has been likened to that of a cat in distress.

INBORN ERRORS OF METABOLISM: MOLECULAR DISEASE

During the first quarter of the twentieth century Sir Archibald Garrod coined the phrase 'inborn error of metabolism' when suggesting a reason for the occurrence of particular diseases in families — alkaptonuria, for example. Garrod went on to affirm '....the most probable cause is the congenital lack of some particular enzyme, in the presence of which a step is missed and some normal metabolic change fails to be brought about'.

Garrod's concept was later confirmed as a result of extensive mutation studies utilizing the mould *Neurospora grassa,* from which Tatum and Beadle hypothesized the one gene—one enzyme theory (Beadle, 1959). Since the living process is seen to be controlled almost exclusively by enzymes, each of which is either evoked or controlled by a single gene or group of genes, Beadle's hypothesis states, in effect, that an organism has a genotype which can give rise

TABLE 6.5

Some Probable Enzyme Deficiencies in Different Conditions

Condition	Enzyme deficiency
Albinism	Tyrosinase
*Alkaptonuria	Homogentisic acid oxidase
Cystathioninuria	Cystathionine cleavage enzyme
*Galactosaemia	Galactose-1-phosphatase
	Uridyl transferase
	Glucose-6-phosphatase
	Amylo-1, 6-glucosidase
Glycogen storage	Amylo-(1: 4–1: 6)-transglucosidase
Hypophosphatasia	Alkaline phosphatase
Methaemoglobinaemia	Diaphorase 1
*Phenylketonuria	L-phenylalanine hydroxylase
Primaquine sensitivity	Glucose-6-phosphate dehydrogenase
Suxamethonium sensitivity	Serum cholinesterase
Tyrosinosis	p-hydroxy-phenylpyruvic acid oxidase

to an enzyme profile which will determine its basic behavioural pattern; the expression of the interaction of the genotype with the total environment is known as the phenotype. Furthermore, the genotype may be changed by mutation, on the basis of each gene being responsible for the controlling of the synthesis of one enzyme, which is taken as a change occurring at the locus of a gene — thus precipitating a change in the phenotype. If a gene is absent, abnormal or deficient, a corresponding effect will be manifested in the enzyme community — resulting in molecular disease. The implication is then that life is essentially a phenomenon arising out of a relationship between a large and specific molecular population, and not simply the property of a single molecule. In consequence, a molecular situation, or incident motivating an alteration in the specific molecular community which places the system in jeopardy is classified as a disease. Thus, by virtue of the existence of a direct relationship between a gene and the presence of a particular enzyme, it is possible to explain certain biochemical disturbances characteristic of the disease in terms of an obstruction occurring at some stage in the normal events of the metabolic process — the obstruction being directly attributable to the deficiency of a specific enzyme, which in turn can be explained in terms of a single gene substitution. Some examples which are seen to fall into this classical pattern are listed in Table 6.5 — and those asterisked will be among those briefly discussed. (For a more detailed exposition of metabolic disorders *see* Thompson and King, 1964.)

Alkaptonuria

This is a rare condition readily recognized by the characteristic way in which freshly voided urine (at first normal in appearance) rapidly darkens in colour on standing. This change in colour is accelerated by alkalinity, and eventually the urine becomes black — due to the presence of a excess amount of homogentisic acid which is rapidly oxidized to a series of brown and black pigments.

Homogentisic acid is an intermediary product in the normal metabolic pathway in the degradation of two amino acids — phenylalanine and tyrosine. Thus alkaptonuria is a metabolic disorder in which the further oxidation of homogentisic acid to maleylacetoacetic acid is blocked by the deficiency in the enzyme homogentisic acid oxidase (*Figure 6.16*). Genetically there are two types of alkaptonuria. In the most common form of the disease, the affected individual is homozygous for the abnormal gene, thus the disease is determined by a recessive gene, whereas in the rarer form the affected individual is heterozygous.

Phenylketonuria

Inherited as a typical Mendelian recessive character this is a condition which, is generally produced by consanguinous unions. Heterozygotes for the abnormal gene are apparently clinically normal, and can be detected by the phenylalanine tolerance test. Phenylketonuria can be demonstrated by the presence of excessive quantities of phenylpyruvic acid in the urine. Normally phenyl-alanine — which is an essential amino acid, derived either from the dietary or tissue proteins — is irreversibly converted to tyrosine by the enzyme L-phenylalanine hydroxylase. In this disease, however, the mutated gene has resulted in the lack of this vital enzyme needed for the oxidation of phenyl-alanine to tyrosine, with the net result that it becomes oxidized to phenylpyruvic acid which is excreted in the urine. The existence of this block in the

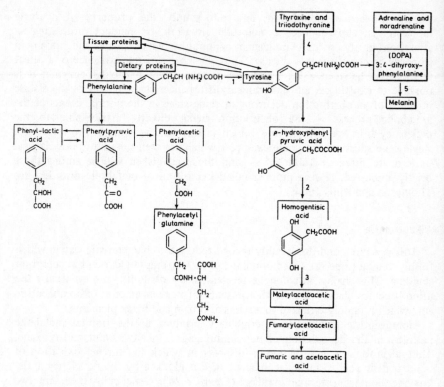

Figure 6.16. Diagram indicating sites of genetic blocks occurring in the metabolism of phenylalanine and tyrosine in man: 1: phenylketonuria 2: tyrosinosis 3: alkaptonuria 4: goitrous cretinism 5: albinism

intermediary metabolism of phenylalanine has been clearly demonstrated by monitoring the plasma tyrosine levels in both normal and phenylketonuric patients following a phenylalanine meal. The graph in *Figure 6.17* shows that in the normal host the phenylalanine leads to a marked rise in the plasma tyrosine level, whereas in the phenylketonuric patient there is no such manifestation. Phenylketonuria is thus the primary action of the gene – the result being the accumulation of phenylpyruvic acid. In such cases the effect is to produce idiocy. It would seem that since phenylalanine is the most prominent abnormal constituent in such cases, this is the causative agent of the mental defectiveness. In fact the precise cause is unknown, since a clear correlation has never been found between IQ and the blood levels of phenylalanine – or, indeed, of any other abnormally increased metabolite (Fuller and Shuman, 1969).

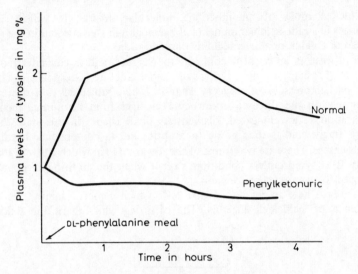

Figure 6.17. Effect of phenylalanine meal on plasma tyrosine level in a normal and in a phenylketonuric patient

Von Giercke's Disease

This is sometimes known as glycogen storage disease, and manifests in a variety of clinical forms depending on the deficient enzyme implicated for degrading glycogen. Usually these patients exhibit hepatomegaly and attacks of hypoglycaemia and ketosis — these being due to a deficiency of the enzyme glucose-6-phosphatase.

Congenital Galactosaemia

This term refers to a situation in which a gene is deficient in controlling the synthesis of the enzyme galactose-1-phosphate uridyl transferase which mediates in the conversion of galactose to glucose. Patients with this disease generally fail to thrive, manifesting mental retardation, severe liver damage and cataracts.

Figure 6.16 indicates the various genetic blocks which can occur in the metabolism of the amino acids, phenylalanine and tyrosine in humans — so precipitating the genetic syndromes: phenylketonuria; tyrosinosis; alkaptonuria; cretinism; and albinism.

Sickle-cell Anaemia

As implied earlier genes determine the structure of proteins by dictating the assembly of amino acids in a specific sequential order. Sickle-cell anaemia is an

221

admirable example of an inherited molecular disease, in which specific differences in amino-acid components of haemoglobin are reflected in the rate of synthesis of an abnormal haemoglobin configuration.

This disease is apparently confined to Negroes, and is characterized by a change in the shape of the erythrocytic superstructure in response to changes in the partial pressure of oxygen. That is, when subjected to a decrease in oxygen tension the diseased erythrocyte changes from its normal biconcave configuration to an elongated, filamentous sickle shape. In this state the cell is more fragile and is thus prone to rupture, and in consequence haemolytic anaemia ensues. There are two forms of the disease: (1) in which a severe anaemia results; (2) a symptomless condition, except when the crythrocyte population is exposed to an anoxic environment.

It is useful here to refer to *Figure 6.18* which depicts a family tree of the inheritance of sickle-cell anaemia. The offspring who inherit the sickle cell

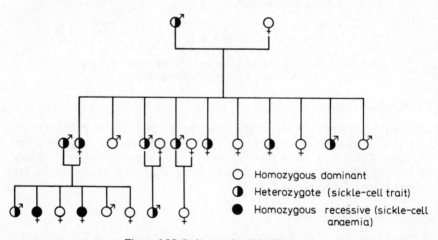

O Homozygous dominant

◑ Heterozygote (sickle-cell trait)

● Homozygous recessive (sickle-cell anaemia)

Figure 6.18. Pedigree of sickle-cell anaemia

character from *both* parents will manifest the severe form of the disease — that is, sickle-cell anaemia. The offspring inheriting the character from only a single parent, on the other hand, will present only a mild form, or the sickle-cell trait. Thus the homozygous recessive individual has severe anaemia, and the heterozygous individual will manifest sickling but only under certain conditions and with no other deleterious symptoms.

In Chapter 2 it was seen that the haemoglobin molecule contains in the region of 600 amino-acid residues arranged on four polypeptide chains designated α and β. Table 6.6 depicts examples of amino-acid substitution occurring in the human Hb molecule. Using a technique called finger printing, which involves partial digestion of the protein by one or more proteolytic enzymes followed by a determination of the pattern of the product peptides on electrophoretic or chromatographic papers, it was found that HbS and HbA differed only in that valine had replaced glutamic acid in the sixth position on the β chain. Thus in sickle-cell anaemia, only abnormal proteins are produced in the homozygous state, and in the heterozygous state both normal and abnormal proteins are synthesized.

TABLE 6.6

Amino Acid Substitution occurring in Human Haemoglobin Molecule

Type of abnormal Hb	No. of residues involved	Normal residue in HbA	Residue in abnormal Hb	Abnormality occurring in α or β
I	16	Lys	Asp	α
C	6	Glu	Lys	β
S	6	Glu	Val	β
E	26	Glu	Lys	β
MBoston	58	His	Tyr	α
GPhiladelphia	68	Asn	Lys	α
GSan Jose	7	Glu	Gly	β

In normality haemoglobin (HbA) consists of two α and β chains possessing 141 and 146 amino acids respectively. The terminal ends of the α chain are valyl–leucyl, whilst in the β chain they are valyl–histidyl–leucyl residues. Other variations, including the one cited, are all reliant on changes occurring in the globin moieties as a result of mutations of the α chains – for example, HbI, P, Q, and D; or of the β chains – for example, HbS, C, D, E, G, J, L, and N of HbA. On the other hand abnormalities can occur as a result of α or β chain synthesis of the HbA molecule being suppressed – as in thalassaemia. This disease exists as does sickle-cell anaemia in two forms, thalassaemia major and minor respectively. Homozygous β thalassaemia, referring to the β polypeptide chain, generally produces the major form, and heterozygous β thalassaemia produces the minor. Also one form of homozygous α thalassaemia has excess β chains which polymerize, forming an unstable tetramer of HbH. In the case of homozygous β thalassaemia, the resultant excess of chains may combine and be precipitated into the cell cytoplasm as Heinz bodies, whereupon these cells haemolyse.

BIBLIOGRAPHY AND REFERENCES

Barr, M. L. and Bertram, E. G. (1949). 'A Morphological Distinction between the Neurones of the Male and Female, and the Behaviour of the Nucleolar Satellite during Accelerated Nucleoprotein Synthesis.' *Nature, Lond.*, **163**, 676.

Beadle, G. W. (1959). 'Genes and Chemical Reactions in Neurospora.' *Science, N.Y.*, **129**, 1715.

Bergsma, D. (Ed.) (1966). Chicago Conference: 'Standardisation in Human Cytogenetics'. *Birth Defects* Original Article Series, Vol 2, No. 2.

Brachet, J. and Mirsky, A. E. (Eds) (1959). *The Cell*, Vols 1–5. New York; Academic Press.

Cairns, J. (1963). 'The Bacterial Chromosome and its Manner of Replication as seen by Autoradiography.' *J. molec. Biol.*, **6**, 208.

Clegg, J. B., Weatherall, D. J., Nakorn, S. N. and Wasi, P. (1968). 'Haemoglobin Synthesis in beta-Thalassaemia.' *Nature, Lond.*, **220**, 664.

Conference (1960). 'A Proposed Standard System of Nomenclature of Human Mitotic Chromosomes.' *Lancet*, **1**, 1063.

– (1963). 'The Normal Human Karyotype.' *Cytogenetics*, **2**, 264.

Coult, D. A. (1966). *Molecules and Cells*. London; Longmans Green.

Crick, F. H. G. (1962). 'The Genetic Code.' *Scient. Am.*, **207**, 66.

– (1963). 'The Recent Excitement in the Coding Problem.' In *Progress in Nucleic Acid Research and Molecular Biology*, vol 1, page 163. Ed. by I. N. Davidson and W. E. Cohn. New York; Academic Press.

– Barnett, L., Brenner, S. and Watts-Tobin, R.J. (1961). 'General Nature of the Genetic Code for Proteins.' *Nature Lond.*, 192, 1227.

Davidson, W. M. and Robertson-Smith, D. (1954). 'A Morphological Sex Difference in Polymorphonuclear Neutrophil Leucocytes.' *Br. med. J.*, 2,6.

Drets, M. E. and Shaw, M. W. (1971). 'Specific Banding Patterns of Human Chromosomes.' *Proc. natn. Acad. Sci. U.S.A.*, **68**, 2073.

Dyke, S. C. (Ed.) (1969). *Recent Advances in Clinical Pathology*, Series V. London; Churchill.

Editorial (1968). 'Molecular Biology Comes of Age.' *Nature, Lond.*, 219, 825.

Ford, C. E. and Hamerton, J. L. (1960). 'The Chromosomes of Man.' *Nature, Lond.*, **178**, 1020.

Fuller, R. N. and Shuman, J. B. (1969). 'Phenylketonuria and Intelligence: Trimodal response to Dietary Treatment.' *Nature, Lond.*, 221, 639.

Jacob, F. and Monod, J. (1961). 'Genetic Regulatory Mechanisms in the Synthesis of Proteins.'*J. mol. Biol.*, 3, 318.

Jukes, T. H. (1962). 'Relations between Mutations and Base Sequences in the Amino Acid Code.' *Proc. natn Acad. Sci. U.S.A.*, **48**, 1809.

– (1963a). 'The Genetic Code.' *Am. Scient.*, **51**, 227.

– (1963b). *Informational Macromolecules*. Ed. by H. J. Vogel, V. Bryson and J. O. Lampen. New York; Academic Press.

– (1969). In *Mammalian Protein Metabolism*, Vol III, page 30. Ed. by H. N. Munro. New York; Academic Press.

Klug, A. (1968). 'Rosalind Franklin and the Discovery of the Structure of DNA.' *Nature, Lond.,* **219**, 808.

Levitt, M. (1969). 'A Detailed Molecular Model for tRNA.' *Nature, Lond.,* **224**, 759.

Mazia, D. (1961a). 'Mitosis and the Physiology of Cell Division.' In *The Cell,* vol 3, page 77. Ed. by J. Brachet and A. E. Mirsky. New York; Academic Press.

— (1961b). 'How Cells Divide.' *Scient. Am.,* **205**, 100.

Mendel, G. (1865). 'Experiments in Plant Hybridisation.' English translation is reprinted in: *Principles of Genetics* (1958) by E. W. Sinnott, et. al. 5th ed. New York; MacGraw-Hill.

Meselson, M. and Stahl, F. W. (1958). 'The Replication of DNA.' *Symp. quant. Biol.,* **23**, 9.

Monod, J., Changeux, J. P. and Jacob, P. (1963). 'Allosteric Proteins and Cellular Control Systems.' *J. molec. Biol.,* **6**, 306.

Moore, K. L. and Barr, M. L. (1955). 'Smears from the Oral Mucosa in the Detection of Chromosomal Sex.' *Lancet,* **2**, 57.

Miller, O. J. (1964). 'The Sex Chromosome Anomalies.' *Am. J. Obstet. Gynec.,* **90** (Suppl.), 1078.

Morgan, T. H. (1919). *The Physical Basis of Heredity.* Philadelphia.

McIntosh, J. R., Hepler, P. K. and Van Wie, D. G. (1969). 'Model for Mitosis.' *Nature, Lond.,* **234**, 659.

Nirenberg, M. W. (1963). 'The Genetic Code.' *Scient. Am.,* **20**, 880.

— and Matthaei, J. H. (1961). 'The Dependence of Cell-free Protein Synthesis in *Escherichia coli* upon Naturally Occurring or Synthetic Polyribonucleotides.' *Proc. natn. Acad. Sci. U.S.A.,* **47**, 1588.

Nowell, P. C. (1960). 'Phytohaemagglutinin. An Initiator of Mitosis in Cultures of Normal Human Leucocytes.' *Cancer Res.,* **20**, 462.

Ochoa, S. (1964). 'Chemical Basis of Heredity: The Genetic Code.' *Experientia,* **20**, 57.

Ohno, S. (1961). 'The Single X Nature of Sex Chromatin.' *Lancet,* **1**, 78.

Perutz, M. F. and Lehmann, H. (1968). 'Molecular Pathology of Human Haemoglobins.' *Nature, Lond.,* **219**, 952.

Pearson, P. L., Bobrow, M. and Vosa, C. G. (1970). 'Technique for Identifying Y Chromosomes in the Human Interphase Nuclei.' *Nature, Lond.,* **226**, 78.

Platt, J. R. (1962). 'A Book Model of Genetic Information Transfer in Cells And Tissues.' In *Horizons in Biochemistry.* Ed. by M. Kasha and B. Pullman. New York; Academic Press.

Pontecorvo, G. (1952). 'Genetic Formulation of Gene Structure and Gene Action.' *Adv. Enzymol.* **13**, 121.

Pritchard, R. H. (1960). *Microbial Genetics.* 10th Symposium Society for General Microbiology. Cambridge; University Press.

Rao, P. N. and Johnson, R. T. (1970). 'Mammalian Cell Fusion: Studies on the Regulation of DNA Synthesis and Mitosis.' *Nature, Lond.,* **225**, 159.

Ris, H. (1957). 'Chromosome Structure.' In *The Chemical Basis of Heredity,* page 23. Ed. by W. D. McElroy and B. Glass. Baltimore; Johns Hopkins Press.

Sinnott, E. W., Dunn, L. C. and Dobzhansky, T. (1958). *Principles of Genetics,* 5th ed. New York; McGraw-Hill.

Tatum, E. L. (1964). 'Genetic Determinants.' *Proc. natn Acad. Sci. U.S.A.,* **51**, 908.

Taylor, A. L. and Thoman, M. S. (1964). 'The Genetic Map of Escherichia coli K-12.' *Genetics, N.Y.,* **50**, 667.

Taylor, J. H. (1959a). 'The Duplication of Chromosomes.' *Scient. Am.,* **198**, 36.

— (1959b). 'Autoradiographic' Studies of Nucleic Acids and Proteins during Meiosis in *Lilium longiflorum. Am. J. Bot.,* **46**, 477.

— Woods, P. S. and Hughes, W. C. (1957). 'The Organisation and Duplication of Chromosomes as Revealed by Autoradiographic Studies using Tritium-labelled Thymidine.' *Proc. natn. Acad. Sci. U.S.A.,* **43**, 122.

Thompson, R. H. S. and King, E. J. (1964). *Biochemical Disorders in Human Disease.* London; Churchill.

— and Wootton, I. D. P. (Eds) (1970). *Biochemical Disorders in Human Disease.* London; Churchill.

Tjio, J. H. and Levan, A. (1956). 'The Chromosome Number of Man.' *Hereditas,* **42**, 1.

— and Puck, T. T. (1958). 'Genetics of Somatic Mammalian Cells. II: Chromosomal Constitution of Cells in Tissue Culture.' *J. exp. Med.* **108**, 259.

— and Whang J. (1962). 'Chromosome Preparations of Bone Marrow Cells without Prior *in vitro* or *in vivo* Colchicine Administration.' *Stain Technol.,* **37**, 17.

Watson, J. D. (1963). 'Involvement of RNA in the Synthesis of Proteins.' *Science, N.Y.,* **140**, 17.

— (1965). *Molecular Biology of the Gene.* New York; Benjamin.

— and Crick, F. H. C. (1953a). 'Molecular Structure of Nucleic acids: A structure of Deoxypentose Nucleic Acids.' *Nature, Lond.,* **171**, 737.

— — (1953b). 'Genetical Implications of the Structure of Deoxyribose Nucleic Acid.' *Nature, Lond.,* **171**, 964.

White, M. J. D. (1967). *The Chromosomes,* 5th ed. London; Methuen.

Woese, C. R. (1965). 'Order in the Genetic Code.' *Proc. nat. Acad. Sci. U.S.A.,* **54**, 71.

Valentine, G. H. (1966). *The Chromosome Disorders: An Introduction for Clinicians.* London; Heinemann.

Yunis, J. J. (1965). *Human Chromosome Methodology.* New York; Academic Press.

Zeuthen, E. (Ed.) (1964). *Synchrony in Cell Division and Growth.* New York; Interscience Publishers.

7−Immunology−Self Versus Non-Self

(Antigens − Structure and nature of immunoglobulins − Antibody formation − Antigen−antibody interaction in vitro − Antigen−antibody interaction in vivo − Mammalian blood groups − Host-parasite relationship

This subsection of Part IV will deal with the phenomenon of recognition of self, the manifestation of which can be likened to the assertion of molecular ego − that is the ability of the molecular community of an organism to recognize a configuration compatible with its own somatic complex ('self'), and reject those configurations which are foreign (non-self) and consequently incompatible. Those mechanisms involved in such a recognition programme are the vital components of the multifactorial science known as Immunology.

Historically the study of immunology stemmed from a consideration of that aspect of microbiology which was primarily concerned with explaining the phenomenon of immunity − that is, the ability of a particular organism either to remain exempt from the harmful influences of such agents as micro-organisms, or to acquire this immune status following exposure to such an agent.

Though it was not until 1876 that Koch proved conclusively that micro-organisms were the causative agents of infectious diseases, the observation of acquired immunity gave rise to the use of cowpox vaccination against smallpox by Jenner in 1798. Later in 1890, von Behring and Kitsato discovered, whilst working with animals infected with *Corynebacterium diphtheriae,* that the sera of these animals possessed a substance capable of neutralizing the *C. diphtheriae* toxin − thus demonstrating for the first time the existence of a substance in a host which would specifically oppose an invading agent, and as a consequence afford the host a measure of protection, in conferring immunity on the somatic system. Consequently the 80 years of immunological research which followed have been directed towards elucidating the immune mechanism, which is formulated following a challenge initiated by the introduction of a foreign substance,

227

or antigen into the body which stimulates in turn the production of a protective substance known as an antibody. Following this, it is apparent that this new immune experience is added to the immunological 'memory' of the organism, such that it can be utilized against a subsequent invasion by the same or a related antigenic species. This memory is established by means of inheritance and intra-uterine and extra-uterine experience. Some aspects and principles which are known regarding the immune mechanism will now be reviewed.

ANTIGENS – A GENERAL CONSIDERATION

In general an animal will not form antibodies against itself, and clearly there is a mechanism in operation whereby an injected foreign substance is differentiated as such by the system of the recipient – so initiating a specific synthesis pathway. It is important at this stage to clarify what is meant by foreign. The implication is that the fundamental requirement of a substance for it to be classified as foreign, and to serve as an antigen, is that it be alien to the antibody-forming tissues of the host and yet not necessarily to the animal as a whole. For though the opening statement was: 'an animal will not form antibodies against itself' – this is a gross generalization and is not strictly true, for auto-antibodies can be produced by injecting an animal with such of its own inaccessible tissues as lens and brain proteins, these sometimes being known as occult antigens (*see* Antibody Formation, page 242).

Antigens are macromolecules which are capable of stimulating the formation of specific antibodies. Though substances which are endowed with this antigenic property vary greatly in nature it would appear that there are two prime requisites determining antigenicity. One is molecular weight, since it would seem that those substances with a molecular weight of less than 10 000 are either non-antigenic or else very weak antigens. It must be stressed, however, that this is by no means a strict criterion, for even glucagon – which is a polypeptide with a molecular weight of only 2 100 – is weakly immunogenic and the effectiveness of this can be enhanced by it being complexed with a protein. The second essential, which is probably more important, is that the substance must possess antigen-determinant groups to which a recipient organism is capable of responding.

Antigens may be resident either extracellularly or intracellularly, but in either case they are capable of stimulating the formation of specific antibodies in an appropriate host. The position of the antigen in relation to the cell structure, and the surrounding tissue environment are also important – not only in respect of the production of antibody but also of the facility with which antibody can reach the cell antigens and thus react with them. Furthermore the type of antigenic species associated with the cells of an individual organism is seen to be dependent solely on the cell type, since tissues and organs vary greatly in their cellular composition. For instance, certain antigens may occur only in the heart, and others only in the liver; such antigens are referred to as organ-specific, and these, in effect, will not only manifest specificity characterizing the individual and the species, but will also represent specialized products of cells and intrinsic components of the cell structure. This is not surprising really since such specificity is merely a reflection of phylogenetic development.

The degree of immunogenicity of an antigen appears to be influenced by the phylogenetic relation of the donor to the recipient. For instance human proteins will stimulate a greater response in guinea-pigs than in the monkey. Hence the greater the phylogenetic differences between individuals, the greater the antigenic disparity resulting.

The proteins are good antigens when introduced into a species other than the one from which they originated. Another class of substance also shown to be antigenic are the polysaccharides. Immunologically speaking purified polysaccharides are peculiar in as much as they will only stimulate antibody formation in certain species and not in others. For example, both guinea-pigs and rabbits will respond to protein antigens, but neither will produce antibodies to purified polysaccharide; on the other hand, mice and human beings will respond to the latter. An important group of polysaccharides are the capsular antigens of the pneumococcus which can be purified from culture filtrates. Not only are they antigenic in man, but also antibodies to these polysaccharides have been shown to protect animals against pneumococcal infections (*see* Chapter 4, page 97 and Chapter 6, page 192). It is important to note here that when the total encapsulated organism – that is, the polysaccharide complexed with protein – is introduced into guinea-pigs and rabbits then antibodies are formed. In the Gram-negative bacteria, where the cell wall lipids are complexed with protein and polysaccharide, it is believed that the lipids serve as antigen-determinant sites (Table 4.1). This is of interest since pure lipids and long hydrocarbon chains appear to be non-antigenic. It is considered that this may be attributable to a lack of structural rigidity and repetitive nature of the structural units within the total molecule.

On the other hand the glycoproteins and the glycopeptides are antigenic; indeed, these substances constitute the major component of an important class of antigen, called the iso-antigens, which are inherited according to Mendelian genetics. Probably the most important members of the iso-antigens are the blood group substances A, B and Rh (*see* Chapter 3, page 82 and later), the specificity of which is determined by the carbohydrate moiety.

The nucleic acids are generally considered to be non-antigenic, but antibody can be formed against DNA, if the molecule is first denatured, complexed with bovine serum albumin (BSA), and then injected into an animal together with an adjuvant. It is believed that the antigenicity of DNA may be masked by the presence of histones resident in the molecular structure: if this is the case, the histones could be considered to act here as an insulator.

Antigen-determinant Sites

On a molecular basis antigens may be classified into two groups by virtue of their inherent spatial configuration which is pre-determined by the structural sub-unit sequence:

(1)   a molecule containing a repetitive sequence of residues constituting a primary structure in which areas of the linear form serve as determinant sites; as, for example, in the fibrous proteins and the polysacchardies;

(2)   a structure possessing a specific sub-unit sequence which is folded and stabilized in a three-dimensional conformation by covalent and non-covalent bonding; as in the globular proteins, for example. Here the primary linear chains are coiled into a right-handed or left-handed helical structure by the formation of covalent peptide bonds (the secondary structure of protein); this is subsequently stabilized in a tertiary configuration by the presence of disulphide bridges. The resultant structure is characterized by antigen-determinant sites consisting of surface amino-acid groups. The reason for the immunogenic property of proteins is not known, though it is thought that each specific antibody possesses a unique sequence of amino acids which folds in a unique three-dimensional shape and that the function of the antigen is thus to select the synthesis of a specific mRNA template that codes for the desired amino-acid sequence – at least according to the selective theory (*see* Antibody Formation Theories, page 242). Studies of synthetic polypeptides have revealed that those structures with an optical laevorotatory (L) configuration are more readily degradated – and as a result antigenic – whereas those with a D-configuration are non-antigenic simply indicating that in the latter instance a metabolic process is not mobilized.

Haptens

A hapten is a substance which will react specifically with an antibody, but will not initiate antibody formation unless first complexed with a suitable carrier substance. For example, the chemical *p*-azosuccinanilate is such a substance which if complexed with sheep globulin – thus forming a hapten-carrier complex – and injected into a rabbit will stimulate the synthesis of antibodies that not only have specificity to sheep globulin but also to *p*-azosuccinanilate. Similarly it was mentioned earlier that some polysaccharides are antigenic, but that their antibody-evoking ability was manifested only in certain species. For example, the pneumococcal capsular polysaccharides were seen to be antigenic in both mice and man, but not in the rabbit. In the latter case, the polysaccharide is behaving as a hapten – that is, if it were first combined with a protein, antibody formation would be induced, capable of combining with the pure polysaccharide and with the polysaccharide-protein complex. This device of attaching another molecule to a carrier substance to obtain an antiserum in which the resultant antibody has some specificity to the introduced group and complex has been utilized by immunochemists to obtain knowledge of the structures of antigen-determinant sites. The following are some examples of reactions used to introduce these haptens into proteins.

Reactions with penicillin. – The highly reactive penicillin molecule will react with either amino or sulphydryl groups of proteins forming a penicilloyl–protein conjugate or penicillenoic acid–protein conjugate respectively.

Dinitrophenyl derivatives. – By reacting dinitrofluorobenzene with the free amino groups of a protein molecule.

Reactions with isothiocyanate and isocyanates. – Here the reaction is also with the free amino group of the protein.

Antigen Fate In Vivo

Antigenic fate is important, as implied earlier, since it will determine the degree of antibody response and the subsequent interaction of antibody with residual antigen. The fate of antigen will be seen to be dependent on a large number of variable factors, of which the following probably have the greatest bearing *in vivo*: (1) the route of entry; (2) the nature of the antigenic material; (3) the ability of the recipient to cope with the invasion.

Route of Entry

The modes of entry can be divided simply into two categories – natural and artificial. With regard to the former, the common portals of entry are the specialized membrane surfaces of the respiratory and intestinal tract. During pregnancy, antigenic exchange occurs between the foetal and maternal circulations in which the placental complex is mediator between the two environments. The term artificial entry implies the deliberate introduction of a foreign antigen into a somatic system by means of intradermal, subcutaneous, intramuscular or intraperitoneal injections or by surgical implantation. For example, in the administration of drugs, plasma or blood or by means of tissue transplantation.

The route of entry will naturally determine the site of localization of antigen by the phagocytic cells of the reticulo-endothelial system. For if the entry route was intravenous then the principal site of localization would be in the fixed cells of the reticulo-endothelial system; similarly, if the antigen were introduced subcutaneously, or by some alternative route, it would either remain at the site of deposition or else localization would occur in the phagocytic cells of those lymph nodes draining the area associated with the site of entry of the antigen. In Table 7.1 an endeavour is made to classify antigenic fate according to species and portal of entry. As the table shows the physical nature of the

231

antigen – that is, whether it is particulate or soluble on entry – becomes a prime factor in deciding its fate. The rate at which phagocytic cells are able to degradate antigen depends entirely on both the nature of the cell and the antigen involved. It is important to note here that irrespective of the site of entry, particulate

TABLE 7.1
Antigen Fate

Nature of antigen	Portal of entry	Principal site of localization
Foreign red cell (a particulate antigen)	Subcutaneous	Phagocytes of local draining lymph node
	Intravenous	Phagocytes of liver and spleen
Bacteria (a particulate antigen)	Intravenous	Phagocytes of the reticulo-endothelial system, and circulating granulocytes
	Subcutaneous	Phagocytes of local draining lymph node
Foreign albumin (soluble antigen)	Subcutaneous	Reticulo-endothelial system and the phagocytes of the local draining lymph node

antigens are exposed to the threat of phagocytosis, since these cells are widely disseminated throughout the somatic tissue complex. Soluble antigens are more difficult to metabolize initially requiring localization. Certain fixed cells of the reticulo-endothelial system have been shown to exhibit pinocytic uptake of soluble antigens.

Following the intravenous injection of a foreign protein into a non-immune animal, there is an initial decline in the plasma antigenic concentration in which the antigen has – by such means as diffusion – become equilibrated with the blood and extravascular compartments. Particulate antigens, such as bacteria, do not exhibit this initial phase of equilibration and metabolic turnover, since they are phagocytosed. Final elimination of residual antigen occurs at the stage when antibody makes its first appearance, this triggering off interaction between these two complementary components and giving rise to primary complexes which become larger as antibody production is increased. This latter period is known as immune clearance (see Antigen-Antibody Reactions, pages 258, 265). In an organism already possessing immune status in regard to the administered antigen, immune clearance would commence immediately (Rhodes and Sorkin, 1964).

The degradation of an antigen is generally expressed in terms of half-life – time required for concentration in serum to drop from a given value to half that value.

STRUCTURE AND NATURE OF IMMUNOGLOBULINS

Tiselius (1937) using an electrophoretic technique separated serum proteins and designated the resultant fractions albumin, α, β and γ globulins. A year later Tiselius and Kabat (1938) demonstrated that antibodies were present in the γ globulin fraction. Utilizing the Graber and Williams immuno-electrophoretic technique (1955) it was at last possible to increase the resolution of the fraction bands, this facilitating the identification of other serum proteins — particularly the globulins with antibody activity. Thus with the impetus of immuno-electrophoresis three families of immunoglobulins were described: IgG, IgM and IgA. Recently two other immunoglobulins have been characterized, IgD (Rowe and Fahey, 1965 a and b) and IgE (Ishizaka and Ishizaka, 1966; Ishizaka, Ishizaka and Hornbrook, 1966).

Ultracentrifugation studies have revealed normal human serum to be composed of three major components each having different sedimentation (svedberg) constants. These are the 4·5S components — which consist primarily of albumin — and the 7S and 19S components. The immunoglobulins reside mainly in the 7S components, which consist predominantly of IgA, IgG and IgD, and the 19S components which largely comprise IgM.

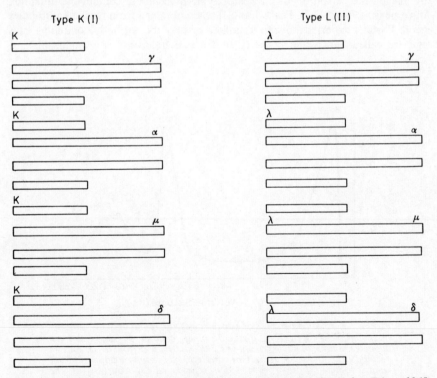

Figure 7.1. Chain composition of 8 known human immunoglobulins (after Fahey, 1965; reproduced by courtesy of the Journal of the American Medical Association)

Structurally the immunoglobulin (IgG, IgA, IgM, IgD and IgE) molecules are composed of two heavy (H) and two light (L) chains. The light chains have a molecular weight somewhere in the region of 22 000, whereas that of the heavy chains is approximately 53 000. The L and H chains are linked by disulphide bridges between the cysteine residues, and there is also weaker interchain bonding. The L chains are a constant feature of the immunoglobulins (including cross-reactivity) – and as such form the basis of group specificity – whereas the H chains are immunoglobulin-specific. There are two types of L chains designated κ chains, and λ-chains. On the other hand the H chains of the individual immunoglobulins IgG, IgA, IgM, IgD and IgE are termed γ (gamma), α (alpha), μ (mu), δ (delta) and ϵ (epsilon) respectively (W.H.O., 1964). In *Figure 7.1* the polypeptide composition of eight known categories of human serum immunoglobulins is depicted diagrammatically.

Employing a proteolytic enzyme (papain) in the presence of cysteine, Porter (1959) demonstrated that the IgG molecule could be split into three segments, by placing the papain digest on a carboxymethylcellulose column and obtaining three peaks (*Figure 7.2*). Peaks I and II when obtained from purified antibodies (as in Porter's experiment) were found to possess the antibody combining sites and are called Fab, while Peak III, unlike Fab fragments is crystallizable and

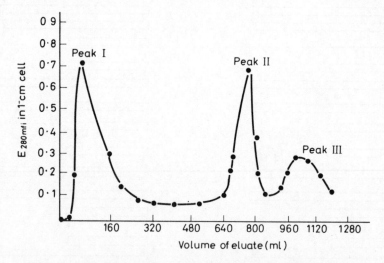

Figure 7.2. Chromatography of papain digest of rabbit IgG on carboxymethylcellulose. Weight of digest 150 mg, column 30 cm x 2·4 cm diameter, volume of mixing chamber 1 200 ml. Gradient from 0·1 M sodium acetate, pH 5·5 to 0·9 M sodium acetate pH 5·5 commencing at 200 ml eluate volume (from Porter, 1959; reproduced by courtesy of the Biochemical Journal)

exhibits no antibody activity. The separated Fab and Fc fragments have 3·5S sedimentation characteristics. Of the total digest prepared with papain two-thirds is Fab and one-third is Fc. Also by virtue of the crystallization character-istics of the Fc fragments it is assumed that this portion of the molecule remains constant in the various species. Degradation studies of the IgG molecule using pepsin (Nisnoff, Wissler and Lipman, 1960) yielded a 5S peptic fragment with a molecular weight in the region of 100 000 which was shown to be a derivative of the two Fab (papain) fragments. Furthermore this 5S derivative — designated F (ab')$_2$, to distinguish it from the papain Fab fragment — was shown to have bivalent antibody activity and, following reduction, it cleaved into univalent fragments having a sedimentation constant of 3·6S. Subsequent oxidation of the Fab fragments yielded a 5S bivalent structure — that is, the original F (ab')$_2$ conformation. From these studies of the IgG molecule it became apparent that architecturally it consists essentially of two monospecific antibody reaction sites.

The first schematic model of the IgG molecule was proposed by Porter (1962); this is probably the most favoured structure, closely adhering to present available experimental data. Alternative structures are those of Edelman and Gally (1962 and 1964) and Tanford (Noelken *et al.*, 1965) — all are featured in *Figure 7.3*.

The Edelman concept is of a cylinder composed of two equal and symmetric halves. Fundamentally the Porter model differs from the Edelman structure only in the location of the L chains, whereas the Tanford structure is seen as an assembly of three sub-units linked at a common spatial nucleus by a disulphide bond which in turn links the two H chains.

As implied earlier the superstructure of the immunoglobulins comprises four polypeptide chains — 2 heavy and 2 light — linked by interchain disulphide bonds. In general every chain has an N-terminal amino acid (the -NH$_2$ group of which is free); in fact the N-terminal ends of the L chains are subject to considerable variation, whereas the C-terminal end-sequences are for the most part invariant.

A characteristic feature of the immunoglobulins is the heterogeneity of the resident polypeptide chains, as shown by carbohydrate analyses; these demon-strate definitive heterogeneity within a single type of chain and a characteristic peptide pattern in the Fd portion of the H chain.

Using antisera developed against various chains it has been possible to demonstrate other instances of heterogeneity due to antigenic differences existing within certain types of chain — these differences being genetically controlled. In the case of the L chains the antigenic determinant sites (Inv sites) are present only on the κ chains and never on the λ chains, and the Inv activity is confined specifically to the Fab portion of the molecule which contains the L chain. The symbol Inv signifies: In for inhibitor and v for the patient contributing the agglutinating sera. There are two possible Inv types, namely Inv^a and Inv^b — thus the Inv κ phenotypes are Inv (a$^+$ + b$^-$), Inv (a$^-$b$^+$) and Inv (a$^+$b$^+$). Hilschmann

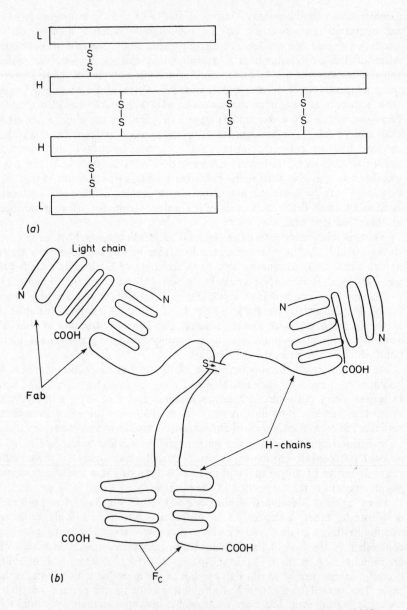

Figure 7.3. Models of the IgG molecule: (a) as proposed by Porter (1962); (b) the Tanford model (after Noelken et al, 1965)

236

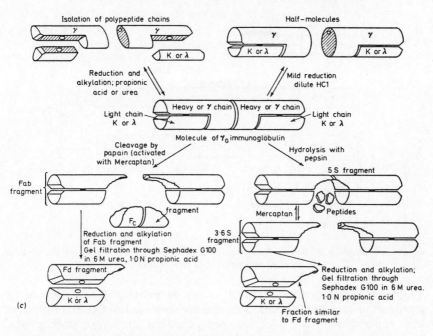

Figure 7.3. (c) Model of 7S IgG molecule. Modification proposed by Fougereau and Edelman (reproduced by courtesy of the Editor of The Journal of Experimental Medicine and The Rockefeller University Press)

and Craig (1965) have shown that in the C-terminal, half remain reasonably constant; in fact the only well documented variation observed in this region among individual proteins is at residue 19I of the human κ chains. Here valine is always present in that position in Inv (b^+), and leucine in Inv (a^+). Thus amino-acid sequence studies have revealed that the light chains have C-terminal sections the resident residue order of which is defined by allotypic and isotypic specificities. The N-terminal half of the light chains is also subject to variation either in isolated positions or in patches affecting two or more residues in a particular area of the chain. It would appear then from recent studies that the N-terminal populace are specific for their clone origin, and that no two chains are identical. The Inv factors found in the light chains of IgG are also demonstrable in IgA and IgM light chains. The diagram in *Figure 7.4* depicts the structural relationship between the light and heavy chains (based on the Porter model), and illustrates those areas which are subject to variation.

With regard to the H chains, heterogeneity of the H polypeptide chains of IgG has been demonstrated. Also four different sub-types of IgG H chains have been shown to exist which have been designated Y2a, Y2b, Y2c and Y2d.

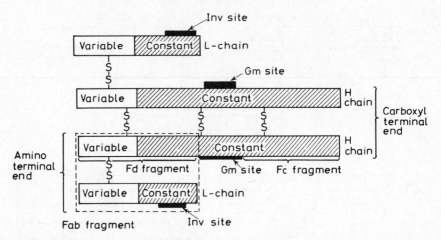

Figure 7.4. Diagram of structure of the IgG antibody molecule, indicating the variable constant regions of the L and H chains

Additional heterogeneity of the H chains is revealed by the presence of genetically determined antigenic determinants known as Gm factors, of which at least 16 have been recognized, these being classified into several allelic groups.

TABLE 7.2

Race	Examples of Gm specifities which behave as units of inheritance						
	a	x	f	b'	c	b^3	b^4
Caucasians	+	–	–	–	–	–	–
	+	+	–	–	–	–	–
	–	–	+	+	–	+	+
Negroes	+	–	–	+	–	+	+
	+	–	–	+	–	–	+
	+	–	–	+	+	–	–
Mongols	+	–	–	–	–	–	–
	+	+	–	–	–	–	–
	+	–	–	–	–	+	–
	+	–	+	+	–	+	+

The distribution of these Gm factors is related to particular peptide patterns of the primary structure of IgG. Furthermore specificities of the Gm system are inherited in certain fixed combinations which vary from race to race (Table 7.2) (Steinberg, 1966; Kunkel *et al.*, 1966). Again anti-Gm factors may occur in the serum of recipients of multiple blood transfusions or injections of IgG, and also as the result of foetal maternal Gm incompatibility.

Comparatively, the immunoglobulins can be visualized as complexes of a basic unit, the monomer consisting of two light and two heavy polypeptide chains – as found for example in the case of IgG. IgA may occur in the form of a monomer, dimer or trimer; IgM, however, is seen as a pentamer unit (*Figure 7.5*).

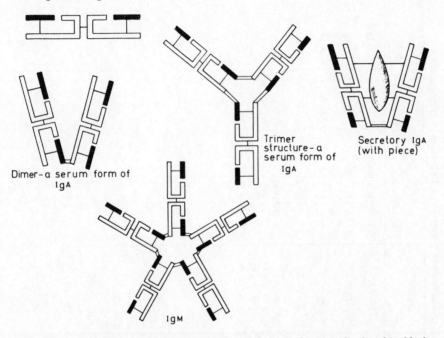

Figure 7.5. Comparative structures of the immunoglobulins (schematized): the white blocks represent the heavy chains and the black, the light chains

The Normal Immunoglobulins

Table 7.3 summarizes the classification of the immunoglobulins and some of their important properties.

IgG Immunoglobulin

The molecular weight of this class of immunoglobulin is in the region of 160 000 and the sedimentation constant is 7S. The class IgG constitutes approximately 70 per cent of the total immunoglobulin population; the

TABLE 7.3

Classification and Properties of the Human Immunoglobulins

Property	IgG	IgA Serum	IgA Secretion	IgM	IgD	IgE
Molecular weight	160 000	170 000–600 000	350 000	900 000	160 000	160 000
Sedimentation constants (S)	7S	7S, 11S, 18S	11S	19S	7S	8S
L-chain type	κ, λ	κ, λ	κ, λ	κ, λ	κ, λ	κ, λ
H-chain type	$\gamma a, \gamma b, \gamma c, \gamma d$	α	α	μ	δ	ϵ
Molecular formulae	$\gamma a_2 \kappa_2$ or $\gamma a_2 \lambda_2$ / $\gamma b_2 \kappa_2$ or $\gamma b_2 \lambda_2$ / $\gamma c_2 \kappa_2$ or $\gamma c_2 \lambda_2$ / $\gamma d_2 \kappa_2$ or $\gamma d_2 \lambda_2$	$(\alpha_2 \kappa_2)n$ or $(\alpha_2 \lambda_2)n$	$\left.\begin{array}{l}(\alpha_2 \kappa_2)n \\ \text{or} \\ (\alpha_2 \lambda_2)n\end{array}\right\} + \text{P}^*$	$(\mu_2 \kappa_2)n$ or $(\mu_2 \lambda_2)n$	$\delta_2 \kappa_2$ or $\delta_2 \lambda_2$	$\epsilon_2 \kappa_2$ or $\epsilon_2 \lambda_2$
Allotypes Gm (H-chain)	+	0	0	0	?	?
Allotypes Inv (L-chain)	+	+	+	+	?	?
Carbohydrate (%)	3	12	12	12		
Active in complement fixation	+	0	0	+		
Placental passage	+	0	0	0	0	
Antibody activity	+	–	–	+		
Normal adult level (mg)	600–1 600	140–420	–	50–190	0·3–4·0	
Synthesis rate (mg/kg/d)	20–40	2·7–55	–	3·2–16·9	0·03–1·49	
Catabolic rate (% per day)	4–7	14–34	–	14–25	18–60	
Presence in CSF	+	+		0		
Skin sensitization Hetero	+	0		0		
Skin sensitization Homo	0			0		

240

carbohydrate content is 2·5 per cent. IgG is not homogeneous, but contains four sub-classes, designated IgG_a; IgG_b; IgG_c and IgG_d. This class contains a large number of antibodies; indeed in the light of its apparent sophistication it is generally considered that — from an evolutionary point of view — IgG seems to have evolved later than IgM. Table 7.4, indicates those antibodies found in association with IgG and the other immunoglobulins.

TABLE 7.4

Depicting Examples of those Antibodies Associated with a Particular Immunoglobulin Class

Type of antibody	IgG	IgA	IgM
Anti-A, B	0	0	+
Immune anti-A, B	+	0	+
Anti-Rh	+	0	0
Anti-Lewis	0	0	+
Cold agglutinins	0	0	+
Le factor	+	0	+
Autoantibodies	+	0	+
Rheumatoid factor	0	0	+
Opsonins	+	0	+
Antitoxins	+	0	0
Complement fixing	+	0	0
Antipneumococcal	+	0	0
Wasserman antibody	0	0	+

IgM Immunoglobulin

Normally some 5—10 per cent of the total serum immunoglobulin content in the human adult consists of IgM. Two sub-classes of IgM have been recognized in human and rabbit serum. These are the largest immunoglobulin molecules possessing a molecular weight of some 900 000 and a sedimentation constant of 19S. For antibodies associated with this class *see* Table 7.4.

IgA Immunoglobulin

Approximately 10 per cent of the total plasma immunoglobulins are normally of the class IgA. Interesting to note is that this class has a variable sedimentation coefficient: 7S, 11S and 18S. Their molecular weights range from 150 000 to 500 000. The 7S variety appear to be composed of two H and L chains, whereas the 11S variety may be dimers of IgA.

Tomasi and Zigelbaum (1963) demonstrated the presence of IgA in large quantities in parotid saliva; as a result of their findings further investigation on other somatic secretions followed, when IgA was found also in colostrum, tears, nasal tracheobronchial and gastro-intestinal (the so-called coproantibodies) secretions. It appears however that the IgA present in these secretions and in the colostrum is different from that found in plasma (Tomasi *et al.,* 1965). Tomasi suggests that secretory IgA is a complex of IgA and a unit which he calls a transport piece which facilitates transport of the complex into the secretions. The transport pieces are now called T-chains, and constitute a polypeptide which becomes attached to the L chains of the IgA structure (Asofsky and Small, 1967). From the above it is apparent that the component T-chain represents a polypeptide chain present in exocrine IgA, but not in serum IgA. It is conjectured that the prime role of secretory IgA in the exterior cavities and membranes is one of defence – that is, it affords these areas a measure of protection against infection.

IgD Immunoglobulin

This is a relatively new immunoglobulin (Rowe and Fahey, 1965b), unrelated antigenically to IgA, IgG and IgM. It is found normally in human plasma in very small concentrations of approximately 30 μg/ml.

IgE Immunoglobulin

It is believed that reagin antibodies belong to this most recently discovered immunoglobulin.

ANTIBODY FORMATION

All vertebrates with the exception of the hagfish are capable of mounting an antibody response – that is, they can synthesize immunoglobulins and reject allografts. It would appear that this ability to produce antibodies is the result of evolutionary sophistication necessitating a system for acquiring a supplementary defence against neoplastic and parasitic disease. Hence our present concept of the immunological phenomenon is one primarily connected with that of somatic protection; indeed, as Burnet (1967) has said: 'an intact immunological system is an essential mechanism for the elimination of malignant cells arising within the somatic complex as a result of mutation'.

Earlier in this chapter the subject of the route of entry of an antigen into the body was raised; this is most important since it has an essential bearing on the general immune response, and the subsequent formation of antibodies in the animal as a whole. Furthermore antigenic potency can be enhanced if the antigen is able to persist in the tissues for a prolonged period of time. This can be achieved by the use of an adjuvant. An adjuvant is a substance which, though not necessarily antigenic in itself, will increase the response to an antigen with which it is injected: for example, *C. diphtheriae* toxoid. This involves the administration of aluminium hydroxide to which the soluble antigen (toxoid) is adsorbed, so facilitating the slow release of antigen into the tissues. The essential features of an adjuvant are thus: (1) retardation of antigenic degradation — thereby permitting prolonged antibody production, which though of a low level is nevertheless effective; and (2) to provoke an inflammatory response in the tissues of the host, which will motivate the migration to the injection site of those cells incriminated in the antibody synthesis mechanism. In addition to the route of administration, the antigen dose is another important factor. Much as with the propagation of an action potential in nerve tissue a threshold level has to be reached in the host's tissues, this being the minimal dose capable of provoking synthesis of antibody. However, once this threshold is exceeded increasing amounts of antigen over a wider range will lead to a proportionately increased response in the host. (In this, it must be remembered, it is unlike action potential propagation.)

The Primary Response

Following an initial exposure to an appropriate antigen (the primary dose) specific antibodies can be detected in the host after a few days. The titre of these rises, and then slowly declines. Actually it is not really possible to generalize in regard to the time of appearance of an antibody, as it will vary according to the antigenic species. For instance, following an injection of erythrocytes, antibodies can be detected as early as 3—4 days later, whereas with bacteria antibodies cannot be detected until some 10—14 days have elapsed.

The Secondary Response

When the antibody levels have decreased following an initial exposure to the primary dose, a subsequent encounter with the same antigenic species will evoke an accelerated secondary response which is characterized by: (1) the lowering of the threshold dose of antigen; (2) a shortening of the lag phase; (3) the achievement of an intensified antibody synthesis level — that is, a rise in titre; and (4) an increase in the persistence of (3) in the host.

The magnitude of the secondary response is dependent largely on the timing of the administration of the secondary dose; ideally this should allow time for the antibody level of the primary response to fall to just below that of detectability, this being known as the optimal period. The primary and secondary responses following immunogenic injections are summarized in *Figure 7.6*.

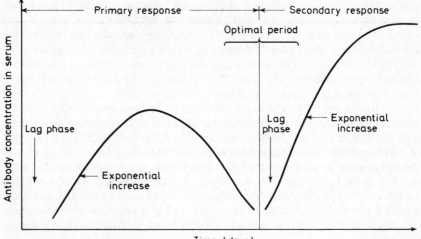

Figure 7.6. Schematic representation of the differential antibody levels achieved in the primary and secondary response following immunogenic injections

Following the introduction of an antigen into the tissues – for example, via the skin, muscle, or subcutaneous tissue the immunogen is transported to the lymph nodes. Histologically the lymph node is composed of lymphoid tissue compartments partitioned by fibrous walls, and within the node body the lymph flows through a fine network of channels called the sinus reticulum. From the circular sinus the lymph percolates slowly through the cortex to the medulla, where it is collected for discharge through the efferent lymphatics. The major functional components of the lymph node structure are the phagocytes lining the sinuses and those cells concerned with antibody formation – that is, the reticulum cells and the cells of the plasmacytic and lymphatic series.

In a resting lymph node approximately 80 per cent of the resident cells are small lymphocytes, and the remainder are of the large variety. Because of their obvious abundance in the lymphoid tissue these were naturally the first cells incriminated in the production of antibody. Until recently, however, there was no direct, conclusive evidence to support this supposition, since it is extremely difficult to distinguish those cells forming antibody (if any) from the other cellular community having a similar morphology. However using the Jerne plaque technique (Jerne, Nordin and Henry, 1963) it is possible to detect antibody-forming cells. Fundamentally, the technique consists of preparing dilute suspensions of immune lymphoid cells in agar to which the foreign erythrocytes (those used for immunization) are added. During subsequent incubation those lymphoid cells synthesizing and secreting antibodies will unite with the surrounding erythrocytes, such that, with the addition of complement to the antibody—erythrocyte system, a discrete plaque of haemolysis is produced in the local area concerned. IgG antibodies are less proficient than IgM in producing this *in vitro* effect — thus the Jerne technique is used only for the latter. By using an indirect method however — that is, by adding specific anti-IgG and complement to the agar, IgG antibodies can be detected, for anti-IgG antibodies will bind specifically with those erythrocytes coated with anti-erythrocytic IgG antibodies. Another means of identifying antibody-forming cells is to employ an immunofluorescent staining technique (Coons, Leduc and Connolly, 1955). The principle of this is: those cells forming antibody (anti-A) are identified by treating them with antigen A; these are then subjected to fluorescent labelled antibody to A when those cells which contain anti-A will bind specifically with them.

As already mentioned the lymph node contains a variety of morphological cells of which by far the most significant is the plasma cell. This, since: (1) as lymphoid tissue becomes hyperplastic in response to an antigenic stimulus, so plasma cells are seen to become more evident in the cell-aggregate population, particularly during the height of the antibody production; (2) the amount of antibody produced compares proportionately with the plasma cell content of the lymph nodes; (3) also in a diseased host (that is, one having hypergamma-globulinaemia), plasma cells are a predominant feature, and as implied in (2) the immunoglobulin concentration is elevated proportionally; (4) finally, plasma cells are virtually non-existent in those hosts unable to synthesize immuno-globulins (as, for example, is the case in agammaglobulinaemia).

The question is from what precursor cells do these plasma cells arise? Following an immunogenic challenge, the number of immature plasma cells is seen to increase. It is conjectured that in fact they arise from a particular species of the small lymphocyte population. Each immature plasma cell (plasmablast) divides, forming new progeny cells, this resulting in successive cytoplasmic differentiation. The development of the plasma cell takes at least eight cell generations, extending over a period of some five days. The prominent feature of the mature plasma cell is its extensive endoplasmic reticulum complex. Hence by the fifth day the adult plasma cells are rapidly secreting specific antibody. Mature plasma cells do not divide. Furthermore their life-span varies in duration from a few days to several months. Those plasma cells producing IgM are believed to have a short life-span.

Recent work seems to suggest that there are two distinct populations of lymphocytes — that is, the short-lived and the long-lived. It is believed that the long-lived type are the memory cells maintaining somatic integrity, and thus responsible not only for the secondary response phenomena, but also for

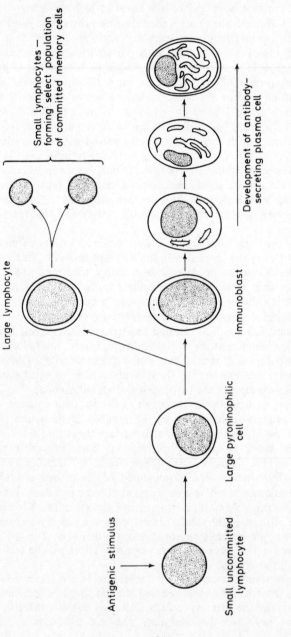

Figure 7.7. Schematic representation of possible lymphocytic metamorphosis following immunogenic stimulation. Note in developing plasma cell the extensive endoplasmic reticulum

long-term immunity status. The general hypothesis is that initially a small uncommitted lymphocyte cell type transforms as a result of an immunogenic stimulus into a large proliferating pyroninophilic cell: that is, a cell capable of dividing into smaller committed lymphocytes or memory cells and immunoblasts which subsequently mature into an antibody-secreting plasma cell. Thus this programmed lymphocyte is able either to divide asymmetrically into: (1) a plasma cell capable of synthesizing a specific antibody; or to (2) differentiate back into a resting lymphocyte which, in effect, constitutes a 'slave' which becomes responsible for the perpetuation of the initial response memory. A schematic representation of the possible lymphocytic metamorphosis following immunogenic stimulation is shown in *Figure 7.7*.

It would appear that the macrophage plays an integral role in the process of antibody formation, for fundamentally it is seen readily to take up antigenic material; furthermore, if the phagocytic activity of such cells is blocked or reduced in some way, then a reciprocating depression of antibody formation is precipitated. In conjunction with the activity of such cells account must be taken of the presence of other immunity substances such as antibody or complement (which functions as an opsonin) which not only motivate these cells, but render susceptible to phagocytosis those foreign particles which otherwise would resist ingestion. From slowly accumulating experimental evidence it seems likely that the macrophage processes the ingested antigenic material in some particular fashion and as a result produces a specific messenger or transfer substance capable of endowing specific cells in the lymphoid series with the potential to produce antibodies (*see* Delayed Hypersensitivity, page 270). Electron micrographs have revealed the existence of cytoplasmic bridges which link the macrophage cells with lymphocytes apparently attracted to these cells containing specific phagocytosed material. It is tempting to conjecture that these cytoplasmic bridges facilitate the transfer of this messenger substance with the potential of dictating a code for specific determinant sites on antibody molecules subsequently synthesized. But it may be that actual contact between the macrophage and the special lymphoid cells is not an absolute criteria. Certainly it appears that the macrophage does act as a vehicle for ingested antigenic material and that it possibly produces a messenger substance. It must be noted here that the theory that these phagocytic cells may possess the potential to metamorphose into antibody-forming plasma cells, has not been entirely dismissed — though present opinion is against this.

At a molecular level the biosynthesis of antibodies (*in vitro*) has established that there is a marked increase in the rate of RNA and DNA synthesis. Further evidence suggests also that the light polypeptide chains of the immunoglobulins are made on small polyribosomes (194S) which then migrate to larger polyribosomes (250S) where the H chains are formed. The synthesis of the L chains is more rapid than that of the H chains — hence there is an excess of L chains which are then secreted into the surrounding environment as free chains.

These excess chains are excreted in the urine. It is believed that the final molecule is assembled via the attachment of free L chains to the polyribosome-bound H chain, these then being released (Scharff *et al.*, 1966). Naturally the synthesis and secretion of antibody is an energy-dependent process. The synthesis rate of the L and H chains varies greatly, particularly when associated with neoplastic conditions. For instance in myeloma the L chains are produced four times as fast as the H chains – indeed in some conditions only the L chains are synthesized. In some human myelomas in which the synthesis of L chains is in substantial excess, they can be detected in urine as Bence-Jones protein.

It is obvious that a molecular homoeostatic control of immunoglobulin synthesis exists, but unfortunately very little is known as to its precise nature. Naturally, however, such a regulatory mechanism involves a specific relationship between catabolism and anabolism. It has been observed that normally the first antibody to appear after antigenization is IgM followed by IgG; as the level of the latter rises so the concentration of IgM diminishes. This is possibly due to the existence of some feedback suppression mechanism coming into play. Furthermore it has been demonstrated that if a specific IgG antibody is administered early on in the primary response period the production of IgM antibody is suppressed; similarly an opposite effect can be precipitated late in the primary response if IgM is administered, as this will result in an increased production of IgG. But whatever mechanisms are involved here, or in other situations related to the immune system, the normal ranges of immunoglobulins are remarkably constant. (For a current review of regulatory mechanisms in antibody synthesis *see* Moller, 1969).

Phylogenetic studies imply that evolution in vertebrates with regard to immunological status is related to the appearance of the thymus and other lympho-epithelial structures. As already mentioned the hagfish (a cyclostome – the most primitive class of vertebrate) does not exhibit any response to an immunological challenge. The Lamprey on the other hand, which is a related cyclostome, appears to possess vestigial lympho-epithelial structures in its gills, which it is conjectured represent a proto-thymus, and as a consequence it is able to mount a weak antibody response to an immunological challenge.

In the higher vertebrates, the ability to form antibodies is a post-foetal capacity. Recent studies have shown that subsequent maturation of lymphoid tissue such as the lymph nodes, spleen and other sites in the reticulo-endothelial system, and their eventual acquisition of an immunological status is due largely to the presence and influence of the thymus. In birds an additional structure is present – the bursa of Fabricius. During embryogenesis the thymus is the first organ to manifest lymphoid elements and appears to be derived from the primitive endodermal layer. The mature thymus is composed of epithelial elements and cells morphologically resembling lymphocytes. It would appear that these thymic lymphocytes form part of the lymphocytic circulation. The precise nature of their migration is unknown; however, it is reasonable to assert that the subsequent cellular interaction resulting from the migration of cells between the thymus, bone marrow and spleen is vital to the establishment of immunological maturation. The probable general migratory routes of the somatic lymphocytic population are indicated in *Figure 7.8.*

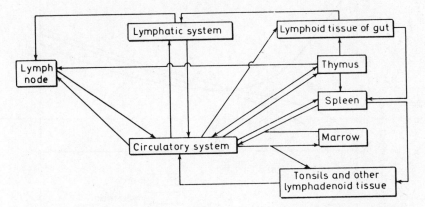

Figure 7.8. A much-simplified scheme of probable migratory routes of mononuclear cells

Late in foetal life the thymus secretes hormones motivating, so it is believed, the maturation of other lympho-epithelial structures. If the thymus is removed at birth lymphoid development is curtailed, and immunological cripples are formed in result. However if the thymus is transplanted into thymectomized neonates, or they are transfused with lymph node, spleen and thymus cells then immunological competence is restored. Also in thymectomized neonates the ability to reject allografts is depressed and, furthermore, the capacity to develop delayed hypersensitivity is more markedly depressed than is the capacity to form humoral antibody.

In normal adult life the thymus slowly becomes atrophied in response to exposure to adrenal cortical steroids. Hence the present concept regarding the role of the thymus is that it appears to regulate the maturation of the progenitors of antibody-forming cells, via the coninuance of thymic hormone secretion.

In birds immunological maturation is achieved not only by the presence of a thymus, but by another lympho-epithelial structure already referred to — the bursa of Fabricius. In man the counterpart of the bursa of Fabricius is believed to be the tonsils, the lympho-adenoidal tissue of the nasopharynx and possibly the lymphoid tissue of the gut. Here the thymus is responsible for the production of humoral antibodies while the bursa is concerned with cellular immunity.

Immunological maturation in mammals commences in neonatal life. Usually synthesis of IgG begins some 4–6 weeks after birth, whereas that of IgA and IgM begins 2–3 weeks after birth and at birth, respectively. This immunological

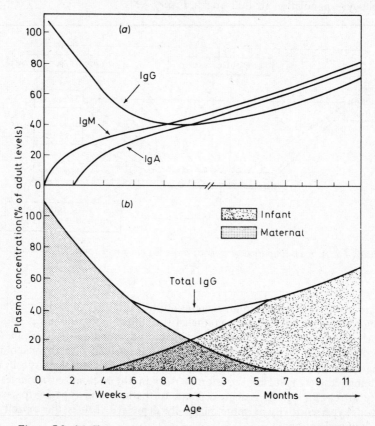

Figure 7.9. (a) Changing concentrations in plasma of the major classes of immunoglobulins in the human infant; (b) disappearance of mother's IgG, and the appearance of the infant's IgG during the first 6 months of life. (From Gitlin, 1964; reproduced by courtesy of Pediatrics, Springfield)

immaturity at birth is because of the limited exposure of the foetus to foreign antigens since, prior to birth, most animal species are protected by a barrier of investing embryonic membranes. If and when a foetus is challenged with an immunological stimulus the principal antibody formed is IgM. In mammals, maternal antibodies can reach the foetus via the placental complex. IgA, IgD, IgE and IgM do not normally pass across the human placenta, whereas IgG is able to pass freely at about the fourth month of foetal life. After birth, maternal antibodies reach the newborn via the colostrum which is rich in IgA and IgG. The IgA is not absorbed but remains in the intestine thereby affording additional protection against infection during the first few weeks of extra-uterine existence, prior to the development of the neonate's own protective antibodies. These are fortified by the presence of IgG which is of prime importance in anti-microbial defence. The changing levels of the major classes of immunoglobulins in the human newly born are depicted graphically in *Figure 7.9.*

Immunological Tolerance

A primary dose of antigen will normally result in the formation of antibodies but if an excessive dose is administered this can result in the suppression of the normal immune mechanism causing what has been termed 'antigenic paralysis', but is now more commonly referred to as immunological tolerance. This induced condition of tolerance can exist for a variable period, the length of which is dependent on the persistence of the particular antigen in the host's tissues. It has been found that antibody production can be inhibited by antigens in other ways: if skin is grafted from one area of the body to another (autograft) then these tissues are readily accepted and will, if transplanted into a suitable vascular bed, survive for an indefinite period of time. Tissue on the other hand grafted from one animal to an unrelated one (homograft) will generally survive for 7—10 days, during which time initial development of a blood supply is established between the graft and the bed, whereupon the graft site becomes inflamed and infiltrated with mononuclear cells. The homograft exhibits advancing peripheral necrosis culminating in the total rejection of the graft. In fact experimentation has shown that a recipient who rejects a graft from an unrelated donor will manifest an accelerated rejection from a second graft, thus being known as a secondary response. Homografts between chimeras, or between animals of given inbred lines, are seen to survive as if they were autografts. This acquisition of immunological tolerance to the homografts, as in the case of bovine chimeras, has arisen from exposure to antigens during foetal life, such that they were recognized as 'self', and as a consequence antibody formation or some other immunological reaction to them did not occur. Medawar and his colleagues (Billingham, Brent and Medawar, 1956) were the first to demonstrate artificially acquired tolerance; the demonstration was devised to test Burnet's hypothesis that tolerance to foreign antigens could be induced by exposure to antigens during foetal or neonatal life. Using mice as experimental animals Medawar injected the foetus *in utero* with lymphoid cells. On reaching maturity these animals were shown to be tolerant to skin grafts from the original donor of the injected lymphoid cell suspension. Similar results were achieved in chickens. Furthermore, tolerance between heterologous species could be produced by parabiosis of turkey and hen, or duck and hen eggs (Billingham, Brent and

Medawar, 1956). Parabiosis is a technique in which two eggs are joined together so allowing the two individual blood circulations to come into contact with one another (Häsek, Lengerovà and Vojtískocá, 1962). Thus the introduction of antigen into foetal animals brings about a failure in the host to recognize the antigen as foreign; instead, when challenged at a later time the antigen is regarded as 'self'. The age of the host at which tolerance can be induced varies: in some it is possible only *in utero,* whereas in others it can be induced shortly after birth or hatching. In the same way immunological competence can be restored to a thymectomized neonate following transplantation of intact thymus or transfusion of lymphoid cells of a normal adult animal: thus it has been possible to produce chimeras experimentally in adult animals by irradiating them with x-rays, and then transfusing them with lymphoid cell from another strain. Chimeras have been produced with rat cells in irradiated mice.

Graft Versus Host Response (GVH) or Runt's Disease

This is a disease precipitated by the transplantation of foreign tissue into newborn animals containing immunological competent cells obtained from the spleen, lymph nodes or thymus of an unrelated adult of a homologous species – that is an allogenic cell transplant. This tissue transplant will be attacked from two directions: (1) by the host-versus-graft (HVG) reaction; and (2) by the graft-versus-host (GVH) reaction. The GVH reaction in the immunologically immature newborn gives rise to a syndrome in growing mice called Runt's disease: this is characterized by emaciation and a failure of the animal to thrive, together with hepatosplenomegaly, and general lymphoid swelling followed by irreversible atrophy and thymic depletion.

This syndrome was first described in 1957 by Billingham and Brent independently, and also by Simonsen (*see* Billingham, 1963; Brent, 1967; Simonsen, 1957). The true nature of the GVH reaction is still not clear, but it seems to be caused by a cellular type of mechanism exerted by the donor cells on the recipient's lymphoid tissues – the thymus, bone marrow, spleen and lymph nodes – where the transfused cells colonize and subsequently proliferate. Furthermore the principal causes of death of the host exhibiting Runt's disease appear to be infection and the development of an auto-immune-like state. If, on the other hand, these immunologically competent cells are administered to an immunologically mature recipient the GVH reaction fails, the cells of the graft being overwhelmed and rejected without injury to the host: this constitutes the HVG response.

With regard to the HVG response, those cells mediating in the rejection phenomenon are the lymphocytes (*see* Delayed Hypersensitivity on page 270). Here the host's lymphocytes come into contact with the foreign antigens of the graft, and it is believed that this contact initiates the formation of a memory image of the foreign antigen. These cells are then carried to regional lymph node centres, whereupon the newly immuno-committed cells multiply, producing a progeny programmed to recognize the invader. Subsequently these cells return to the graft site via passive transport ultimately mediating the rejection of the graft.

Factors Suppressing Antibody Response

Suppression of the immune response can be achieved by the use of chemical agents, x-rays and antilymphocytic serum. Such agents are of importance not only for enhancing the survival of foreign tissue grafts but also for controlling auto-immune disease. So far, the major disadvantage encountered in the use of immunosuppressive agents is their lack of selectivity, but fortunately there are a few which are sufficiently selective to be of clinical use.

Irradiation has been used widely with varying success. Total body irradiation with sublethal doses of x-rays (400–500 r) can suppress immunological activity for up to two weeks, after which time regeneration of the lymphoid tissue occurs. Hence if an immunogen is administered shortly after irradiation, antibody production is delayed until the capacity for mitotic activity is recovered. It has been found in such animals that a second challenge with antigen fails to provoke a secondary response, which rather suggests that the initial irradiation dose prevented the development of the immunocompetent cells required for the secondary response. In fact the effect of the radiation is dependent on the timing of the primary antigen dose – ideally, the most effective times of administration are shortly after and up to 24 hours following irradiation. If, however, the primary dose is given prior to irradiation, thus stimulating some cellular proliferation, then these cells continue their differentiation, and eventually give rise to antibodies. When lethal doses of x-rays are administered to achieve complete immunosuppression, a transfusion with bone marrow is required to replace the destroyed lymphoid tissue of the bone marrow in the host. Though these grafted cells restore defence against bacterial invasion, they unfortunately mount a GVH reaction which has to be checked by the application of supplementary immunosuppressive agents.

Large doses of cortisone and adrenocorticotrophin (ACTH) are seen to have a profound effect on lymphoid tissue and, as a consequence, are capable of suppressing many different kinds of immune reactions. Other immunosuppressive agents are:

The alkylating agents. – These are sometimes called the radiomimetic drugs, since their biological effect resembles that of irradiation. Examples are nitrogen mustard and cyclophosphamide. Such drugs as these cause extensive destruction of the lymphocyte population.

The antimetabolites. − These include the purine analogs such as 6-mercaptopurine and azathioprine; the pyrimidine anologs such as 5-fluorouracil and 5-bromo-deoxyuridine; and the folic acid antagonists such as amethopterin − all of which inhibit normal DNA and RNA synthesis by blocking the formation of essential precursor substances. As with irradiation the antimetabolites are less effective in blocking the secondary response.

Antimicrobial drugs. − Chloramphenicol, administered in high doses, can inhibit a primary response. Experimentally actinomycin D, mitomycin, and puromycin have all been used, but because of their gross toxicity they are of limited clinical use.

Antilymphocytic serum (ALS). − This is an immune serum produced against lymphoid cells of another species: for instance, horse antihuman lymphocyte serum, which is produced by injecting human cadaver spleen or thymus cells into a horse. The precise action mechanism of ALS is still shrouded in mystery, but accumulating evidence suggests that ALS is the least toxic of the immunosuppressive agents at present available. ALS is known to accomplish, among other things, the following:

(*a*) abolition of immunological memory of lymphocytes, thus rendering them incompetent;

(*b*) suppression of delayed sensitivity and cellular immunity mechanisms;

(*c*) suppression of production and release of antibody.

Disordered Immunoglobulin Synthesis

Fundamentally these disorders can be classified into two broad categories characterized by either a deficiency or an increase in immunoglobulin concentrations.

Congenital Agammaglobulinaemia (AGG)

This is sometimes known as Bruton's type after the writer who first described the syndrome, which is a sex-linked congenital disease of male infants (Bruton, 1952). The condition is characterized by defective production of IgG, IgA and IgM. In fact IgG is always below 100 mg/100 ml and IgA and IgM are either absent or else less than 1 per cent of the normal value. Such infants form little or no humoral antibody and as a consequence there is an increased susceptibility to infection unless protection is afforded by an antibiotic umbrella. The organisms generally involved are *Haemophilus influenzae, Streptococcus pneumoniae* and *Staphylococcus pyogenes.* Despite this apparent handicap, the prognosis of these AGG types has improved with the advent of regular gamma globulin therapy.

Swiss Type Agammaglobulinaemia (Hereditary Thymic Aplasia)

This is a fatal condition associated with faulty embryonic development of the thymus, which involves either of the sexes (dependent on autosomal recessive characters) or males only (dependent on an X character). Characteristically the affected infants fail to thrive, and exhibit a history of recurrent or persistent infections. Also tonsillar and adenoid tissue is poorly developed and lymph nodes and thymus are aplastic, or hypoplastic. The immunoglobulins involved are IgA, IgG and IgM and they can either be absent altogether or present in depressed concentrations. Hence this condition presents a variable immunoglobulin profile.

Idiopathic (Primary) Acquired Agammaglobulinaemia

This is a disease capable of manifesting itself at any age, and appearing in both sexes. The immunoglobulins are depressed but never to the extent of those encountered in the congenital forms. Also the causative mechanism is unknown, though workers have endeavoured to show a relationship existing between such cases and the incidence of familial immunological abnormalities such as Lupus erythematosus, thrombocytopenia and haemolytic anaemia. To date cytogenetic studies have produced negative conclusions as to its aetiology.

Secondary Acquired Agammaglobulinaemias

These are conditions manifesting depressed immunoglobulin levels due to neoplasias of the lympho-reticular tissues, and protein-losing enteropathy, nephrosis and other hypercatabolic states.

The Dysgammaglobulinaemias

These are a complex group of diseases associated with changes occurring in one or more of the four immunoglobulin classes. Four types will be defined.

Type I. – This is characterized by an elevated IgM level, and diminished IgG and IgA. The main clinical features are infections of the respiratory tract, progressing to bronchiectasis; lymphoadenopathy and hepatosplenomegaly; haemolytic anaemia, thrombocytopenia and leucopenia manifested in some cases. The prognosis despite this is favourable, though there is a tendency for the disease to become malignant following the intervention of the auto-immune phenomenon.

Type II. – Here infants and adults can be involved, and in adults a sprue-like syndrome develops. IgA and IgG are absent and though IgG may be quantitatively normal it is considered to be non-functional.

Type III. – The total immunoglobulin is normal in this type, but here IgA is either absent or greatly diminished although generally IgG and IgM are quantitatively normal. Both males and females are effected in early childhood. Of those cases manifesting IgA abnormalities 80 per cent have been found to have hereditary ataxia telangiectasis. Furthermore such patients are also susceptible to sinobronchopulmonary infection. The prognosis here is poor, with death occurring in the teens or early adulthood.

Wiskott–Aldrich Syndrome. – This condition affects male infants although there is no consistent immunoglobulin pattern. Clinical features include recurrent pyogenic pneumonias and otitis media. Death generally occurs in late childhood as a result of overwhelming bacterial or disseminated viral infection.

Hypergammaglobulinaemias

Here either one or more of the classes of immunoglobulin may be elevated, and such states are often associated with neoplasma, and the 'collagen' diseases which are predominantly auto-immune in nature.

Theories of Antibody Formation

At present there are three major schools of thought regarding the nature of the mechanism of antibody formation. They are respectively the Germ-line, Instruction and Selection school, of which by far the most universally accepted is the latter.

Historically, the first theory proposed to account for antibody formation was that of Ehrlich (1900). Essentially Ehrlich postulated a selective theory, in which he visualized the surface of the cell soma as being covered by receptor sites which he called side chains, and which he suggested possessed specific configurations capable of combining selectively with particular foreign material. His theory stated further that as a result of antigen–antibody combination, the antigen provoked the production and shedding of these excess side chains which constituted an antibody pool.

By the late thirties Ehrlich's selective hypothesis had become redundant by reason of accumulation of experimental data suggesting that antigenic diversity was so great that it seemed impossible that a lymphoid cell could accommodate the vast potential of possible antigens existing in the environment. As a result other theories were formulated in an attempt to account for this enormous potential number of distinct antigenic classes to which an animal is capable of mounting an immune response. The impetus of this gave rise to the Instruction school the supporters of which believed that by some means the antigen entered and subsequently instructed the antibody-forming cell to produce a specific antibody. In effect the instructionists were saying that the antigen acted as a

256

template which dictated the folding of a nascent immunoglobulin chain, such that its resultant configuration was complementary to the antigen determinant sites (Pauling, 1940). Thus fundamentally the instructionists assumed that antibody specificity was determined by the presence of the antigen at the site of protein synthesis, rather than by the resident amino-acid sequence, which has been shown to be the case. More recently it has been established that antibody molecules can be unfolded, and then re-folded to reconstruct the same specific molecule in the absence of the antigen, whereas according to the Instruction theorists the latter was essential.

Accordingly Jerne (1955) proposed in his paper entitled 'The Natural Selection Theory of Antibody Formation', that antibody-forming cells produce spontaneously a variety of potential antibodies, without prior antigenic stimulation, and that they are able to react with an antigen bearing a complementary configuration, the confrontation of which stimulates the production of specific antibody. As a result of this paper Burnet (1959) formulated the modified Clonal Selection Theory, which to date is probably the most favoured of the theories advanced by immunologists in explaining the majority of the phenomena surrounding the formation of antibody. Burnet's theory states that an organism is capable of elaborating a large number of different clones of inducible cells the progenitors of which are acquired by random somatic mutation. As a result of the introduction of an antigen into the body, selective reaction with a specific cell clone occurs stimulating or inducing it to proliferate and become 'primed'. Further contact with this antigenic species stimulates the 'primed' cell to undergo intensified proliferation and differentiation resulting in the secretion of specific antibody. Also the theory stipulates that only a small number of structural genes from immunoglobulins need be transmitted by germ lines since — as a result of somatic mutation — sufficient diversity would occur to satisfy the amount of genetic information required to code for the vast number of antibody species of an organism need synthesize in relation to the huge antigen spectrum present in the external environment.

Hence a clone represents a unique gene assembly for immunoglobulins the number and diversity of which will be vast. Burnet also postulates the existence of thymus censorship, this doctrine inferring that the thymus eliminates cells from the somatic system the progeny of which would react against 'self'. According to this theory, these 'forbidden' clones of mutant cells arise in adult life, somehow escaping destruction or repression and hence accounting for the formation of auto-antibodies. With regard to the question of tolerance, the clonal theory states that it is more readily produced *in utero* since inducible cells in the embryo are probably more easily repressed by antigen, and also they are less numerous than in the extra-uterine environment. The theory thus explains, and is compatible with, the following immune manifestations; (1) the induction by an antigen of a specific antibody; (2) one cell one antibody; (3) the formation of a 'memory' cell population, which is required for the anamnestic response;

(4) tolerance to accessible self antigens; (5) the formation of an auto-antibody against inaccessible antigen; (6) the ability to induce artificial states of tolerance; (7) that antibody production is genetically determined.

By contrast the germ line theories essentially maintain that the structural genes dictating antibody formation are transmitted from germ cells to somatic cells, which assumes that each inducible cell carries all the genetic information for producing antibodies, and also that each inducible cell is therefore able to respond to two or more antigens.

SOME FEATURES OF ANTIGEN–ANTIBODY INTERACTION IN VITRO

Equilibrium Dialysis

This dialysis is a method employed for acquiring information for measuring an association constant which, in turn, is a measure of interaction between small molecules and macromolecules. That is, for example, the interaction of a low molecular weight hapten with an antibody, which can be expressed as follows:

$$H + A \rightleftharpoons HA$$

where H, represents the hapten and A the antibody. This reaction can be expressed as an equilibrium constant:

$$[HA]/[H][A] = K \ldots (1)$$

Essentially equilibrium dialysis is a system consisting of an antibody and hapten solutions separated by a dialysis membrane – which is permeable to the hapten, but will prevent the passage of antibody. The hapten can be radioactively labelled. Hence when equilibrium is reached, the free hapten concentration is measured on that side not containing antibody. Thus equation (1) may be rewritten so:

$$K = r/(n - r)c \ldots (2)$$

$$\text{or } r/c = Kn - Kr \ldots (3)$$

where r at equilibrium represents the number of antigen molecules bound per antibody molecule at c free concentration of antigen, and n is the maximum number of antigen molecules that can be bound per antibody molecule. Thus from an equilibrium dialysis series in which the antibody concentration remains constant, and the antigen concentration varies, a set of values for r and c can be obtained. By virtue of the heterogeneity of antibody affinity for antigen the expected linear relationship between r/c plotted against r is not found. Thus antigen–antibody association constants will be seen to vary accordingly. Hence it is convenient to define each system as an average association constant (K_o), which exists when one half of the antibody sites are occupied. Thus, by substituting $r = n/2$ in equation (2)

$$K_o = I/c$$

With the formation of an antigen–antibody complex there is a subsequent change in free energy ΔG, which can be calculated from the association constant:

$$\Delta G = -RT.InK_o$$

where ΔG is the standard free energy change, R is the gas constant (1·987 cal/mol-degree), T is the absolute temperature and InK_o, the natural logarithm of the average association constant.

The magnitude of the change in enthalpy or entropy is of considerable value in the understanding of a biochemical reaction.

ΔH may be calculated when K_o has been experimentally determined at two or more temperatures:

$$dInK_o/dt = \Delta H/RT^2$$

therefore:

$$\log_{10} K_2/K_1 = \Delta H (T_2 - T_1)/2·303 \, RT_2.T_1$$

thus:

$$\Delta H = (2·303.R.K_2/K_1)/(1/T_1 - 1/T_2)$$

where K_1 and K_2 are the average association constants at temperatures T_1 and T_2 and the entropy ΔS.

$$\Delta G = \Delta H - T\Delta S$$

The Precipitin Reaction

By comparison with the above which considered the interaction between small univalent antigen and antibody, the precipitin reaction is a situation in which the antigen and antibody are in the soluble form and, as a result of their union, the formed complex becomes insoluble and is subsequently precipitated from solution.

The precipitin reaction takes place in two well-defined stages. The initial association of the antigen with antibody results in the formation of small primary complexes which during the second stage combine to form larger secondary complexes that finally aggregate to form the visible precipitate.

This concept forms the basis of the lattice theory in which the composition of antigen–antibody precipitates is a function of the varying ratios of antigen to antibody. Hence each antigen molecule becomes linked to more than one antibody, which in turn becomes linked to others and so on until aggregate growth reaches a critical size, whereupon spontaneous precipitation occurs. This initial stage is independent of temperature, whereas in the second stage, precipitation is accelerated by increases in temperature. Those forces involved in such a union include van der Waal forces, and hydrogen bonding, which collectively exert an energy level equivalent to some 6 kcal/mol.

In *Figure 7.10* is a graphic representation of a precipitin curve for a monospecific antigen–antibody system. From this it will be seen that in the region of optimal proportion, the supernatants are virtually free of detectable

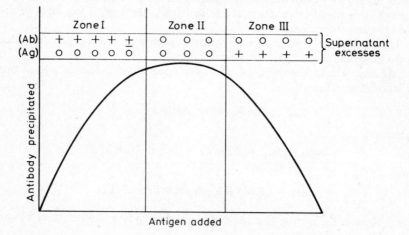

Figure 7.10. Precipitin curve for a monospecific system: Ab: antibody; Ag: antigen; zone I: antibody excess; zone II: equivalence or optimal zone; zone III: antigen excess zone

antigen and antibody whereas on the ascending portion of the curve – that is, the antibody excess zone – the centrifuged supernatants will contain free antibody. Likewise in the descending limb of the curve – known as the antigen excess zone – the supernatant will contain free antigen. The concentration of antibody in an antiserum can be readily determined by utilizing the quantitative precipitin reaction, the principle of which is based on the general concept shown in *Figure 7.10* in which increasing (known) amounts of antigen are added to a constant volume of antisera, and then mixed. The resulting precipitants are then washed and assayed for total protein using either the Biuret or Folin-Ciocalteau reactions. The results are then plotted as shown in *Figure 7.10,* and from this the maximum quantity of antibody which can be precipitated is estimated. Workers such as Heidelberger and Kendall (1935) have endeavoured to simplify the procedure by exploiting the linear relationship existing in the zone of antibody excess, but unfortunately the slope is by no means constant in all antisera of a given specificity, and hence it is of limited value.

The Agglutination Reaction

Contrasting the precipitin with the agglutination reaction the latter occurs with those reactions involving a particulate antigen – the antibody specific for the antigen being located on the cell surface, such that agglutination will occur when added to a cell suspension, by virtue of the fact that a number of bivalent molecules act as bridges uniting adjacent surface antigen determinant sites.

The agglutination reaction is influenced profoundly by the nature of the surrounding environment and the zeta potentials at the cell surface – all of which will affect the normal repulsion forces existing between cells in ordinary media. Ideally, a suitable ionic environment for an agglutination system is one which will dampen the cell's highly negative surface charge, motivating short-range apposition of bivalent antibody molecules with other cells in order to act as specific binding bridges between them. Thus, if the salt concentration is too low, specific agglutination will not occur; optimal salt concentrations are those in the physiological range. Similarly optimal pH values are those approaching neutrality, since when the pH is too low auto-agglutination occurs in the absence of antibody, this sometimes being called acid agglutination.

Prozone is a phenomenon often encountered in agglutination reactions (where there is an excess of antibody), it being a state in which no agglutination occurs in those tubes containing the highest concentrations of antibody. The reason for this is that the antigenic sites have become saturated with antibody, such that the binding effect does not occur. However the prozone phenomenon is not due simply to an antibody excess, but often involves blocking antibodies. Here a portion of the total antibody appears to react with the corresponding antigen in an anomalous fashion such that no agglutination occurs – in fact it is inhibited. This type of antibody is evident in antisera to Brucella.

An antibody capable of manifesting an agglutination reaction is called an agglutinin. It is general practice when employing a semi-quantitative assay, to dilute serially the serum containing antibacterial agglutinins to which is added a constant antigen volume. The highest dilution of this serum which produces agglutination represents an approximate estimate of the relative amounts of antibody in that serum; this is expressed as a reciprocal of the highest dilution, known as the titre. Alternatively the antibody content of a sera containing agglutinins can be estimated by adding a known washed quantity of killed bacteria suspension to a known volume of antisera. The reaction is accelerated by incubating at 37°C or 56°C for a designated period of time. The agglutinate is then centrifuged off, washed with saline and then analysed for total nitrogen (N). The antibody (N) as agglutinin (AbN) is computed by subtracting the quantity of bacterial (N) AgN_1 initially added from the total N found from the washed agglutinated bacteria AgN_2:

$$AgN_2 - AgN_1 = AbN$$

Complement Participation

Complement is used here as a collective term for a group of proteins which can result in irreversible damage to the membranes of susceptible cells, if the antigen either forms part of the membrane's constituents or is adsorbed on to it. The cytotoxic effect of this multifactorial protein system was first described by research workers who made the rather startling observation that if cholera vibrios were injected into the peritoneal cavity of a specifically immunized guinea-pig, these organisms were destroyed by the process of bacteriolysis. Later, Bordet demonstrated this same effect *in vitro,* but found that: (1) if the serum was first heated to 56°C for some minutes this lytic activity was lost – though the antibodies were retained; (2) this activity could be restored by adding fresh normal serum to the system. This thermolabile factor of serum was subsequently called complement, commonly referred to as C'.

When antibodies combine with antigen in the presence of complement the components of C' will participate, and become inactivated – that is, fixed – and thus lose their ability to lyse red cells. This fact is the underlying principle of the complement fixation test. Such tests performed *in vitro* represent a finely balanced and controlled system, which is divided into two – the test system and the indicator system. In system I, known amounts of antisera and antigen are equilibrated in the presence of an experimentally pre-determined quantity of C' – this mixture is then generally incubated at 37°C for 30 minutes, or at 4°C for 12 hours. If antibody is present then the appropriate antigen–antibody complex is formed and fixation of complement takes place. It is important to realize that the test sera (the potential antisera) will possess complement in the system, which has to be inactivated by heating to 56°C for 30 minutes. Thus the only complement existing in the system is that which has deliberately been introduced. At this stage the second system, which constitutes an indicator, is added to the first system in order to determine whether or not complement is present. This indicator system consists of sheep red cells sensitized with anti-sheep cell rabbit serum, often referred to as the haemolysin. In the presence of complement sensitized sheep erythrocytes will lyse – but it has been found that there are several other factors contributing to the lytic reaction, namely the presence of an adequate concentration of magnesium and calcium ions, the value of pH and the temperature. Consequently, if complement is present, haemolysis of the sensitized sheep cells will occur; alternatively if complement is utilized in an antigen–antibody reaction in System I, then haemolysis will not occur this being indicative of a positive result. The complement fixation reaction *in vitro* is summarized in *Figure 7.11.*

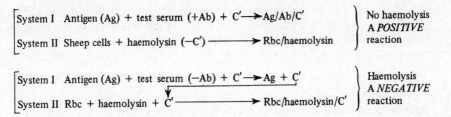

System I   Antigen (Ag) + test serum (+Ab) + C′ ⟶ Ag/Ab/C′   } No haemolysis
                                                                A *POSITIVE*
System II Sheep cells + haemolysin (−C′) ⟶ Rbc/haemolysin   } reaction

System I   Antigen (Ag) + test serum (−Ab) + C′ ⟶ Ag + C′   } Haemolysis
                                                                A *NEGATIVE*
System II Rbc + haemolysin + C′ ⟶ Rbc/haemolysin/C′   } reaction

Figure 7.11. Diagrammatic summary of complement fixation test in vitro

The qualitative complement fixation test just described can be made quantitative simply by diluting the patient's serum used in system I. Complement is generally measured in terms of 50 per cent haemolysis ($C'H_{50}$) – since in this

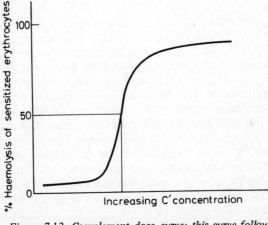

Figure 7.12. Complement dose curve: this curve follows the empirical von Krogh equation $x = K (y/1 - y)(1/n)$ where x is the quantity of C′ added to the system; y is the proportion of cells lysed and n and K are constants. When plotting $C'H_{50}$ it is convenient to plot x against the log $y/1 - y$ hence log $x = log K + 1/n$ log $(y/1 - y)$ in which they intercept at 50 per cent lysis $(y = 1 - y$; log $y/1 - y = 0)$

region there is the greatest change in haemolysis occurring with the smallest changes in C′. To determine this, standard sensitized suspensions of sheep cells are put up against varying dilutions of complement, and the degree of haemolysis is determined spectrophotometrically. In *Figure 7.12* a typical sigmoid curve is shown (the complement dose response curve) which is obtained by plotting the percentage haemolysis against the volume of C' – and the volume giving 50 per cent haemolysis is then read off. For a complete discussion of the complement fixation test see Cruickshank (1960).

The Components of Complement and Reaction Sequence in Immune Haemolysis

As already stated C' is a multifactorial system consisting of at least nine globulin components designated $C'1$, $C'2$, $C'3$, $C'4$, $C'5$, $C'6$, $C'7$, $C'8$ and $C'9$ which appear to function in the completion of those processes initiated by the interaction of antigen and antibody. The majority of the work done in this field has been on human and guinea-pig complement – the designatory correspondence between the two is shown in the following:

Man: $C'1$ $C'2$ $C'3$ $C'4$ $C'5$ $C'6$ $C'7$ $C'8$ $C'9$

Guinea-pig: $C'1$ $C'2$ $C'3c$ $C'4$ $C'3b$ $C'3c$ $C'3f$ $C'3a$ $C'3d$

The following sequential events occur as the result of interaction of erythrocytes (E) with an anti-erythrocytic haemolysin (A) and complement components. Component $C'1$ is a pro-esterase which exists in plasma or serum in the inactive form $C'1p$; this, upon activation, precipitated by the interaction of $C'1q$ and $C'1r$ gives rise to an esterase $C'1a$, in the presence of calcium ions and the antigen–antibody complex. The resultant $C'1a$ in conjunction with the antigen–antibody complex (EA) acts on the $C'4$ component to form a stable complex (EAC'1a, 4). In the next step, which involves the presence of magnesium ions, the bound $C'1a$ complex acts on $C'2$, which is then attached to the bound $C'4$ giving rise to EAC'1a, 4, 2a. As a result of this $C'3$ is taken up giving EAC'1a, 4, 2a, 3. This addition is crucial in the immune reaction, since it not only affects the subsequent sequence, but also causes immune adherence, promoting erythrophagocytosis. The following sequential enzymatic series can be divided into 2 stages: (1) the addition of components $C'5$, $C'6$ and $C'7$ to the preceding complex to produce the stable EAC'1a, 4, 2a, 3, 5, 6, 7; followed by (2) the addition of $C'8$ and $C'9$ which is the final stage resulting in the formation of a damaged site in the erythrocyte membrane causing leakage of intracellular electrolytes and essential metabolites and fluids, culminating in total lysis. The events described here have been greatly simplified; for a more comprehensive survey *see* Wolstenholme (1965). The events of this cascade reaction are summarized in *Figure 7.13*.

It is of passing interest to note that the activity of $C'1$ is inhibited by an inhibitor substance in normality – though apparently this does not interfere with the immune reaction. Patients suffering from a disease known as hereditary angioneurotic oedema have been shown to lack this inhibitor of $C'1$ esterase, such lack causing vascular permeability. Other areas of activity in other immune manifestations are indicated in *Figure 7.13* (Cooper and Fogel, 1967).

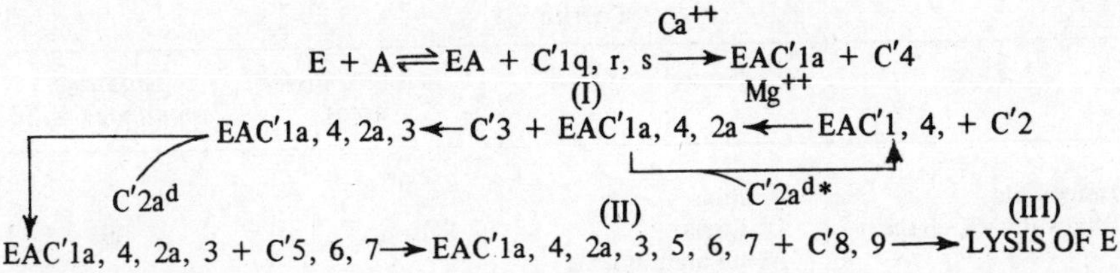

$$E + A \rightleftharpoons EA + C'1q, r, s \xrightarrow{\text{Ca}^{++}} EAC'1a + C'4$$
$$\text{(I)} \qquad \text{Mg}^{++}$$

$$\overset{\displaystyle \Big\downarrow}{} \quad EAC'1a, 4, 2a, 3 \leftarrow C'3 + EAC'1a, 4, 2a \leftarrow EAC'1, 4, + C'2$$

$$C'2a^d \qquad\qquad \text{(II)} \qquad\quad C'2a^{d*} \qquad\qquad\qquad \text{(III)}$$

$$EAC'1a, 4, 2a, 3 + C'5, 6, 7 \rightarrow EAC'1a, 4, 2a, 3, 5, 6, 7 + C'8, 9 \rightarrow \text{LYSIS OF E}$$

*Figure 7.13. Simplification of cascade pathway of the reactions in immune haemolysis. *Indicates that the complex suffers the loss of $C'2a^d$ with decay. E denotes sheep erythrocyte; A anti-sheep rabbit serum. Complex I causes immune adherence and behaves as anaphylactic toxin; II indicated as being responsible for mediating the Arthus phenomenon; III destruction of the erythrocyte*

ANTIGEN–ANTIBODY INTERACTION: IN VIVO

The *in vivo* interactions of antigen and antibody are of profound importance to the organism since they serve principally to protect the host against the possible establishment of foreign agents within the system, which would be detrimental to its integrity. However, often as a result of the combination of antibody with antigen, concomitant tissue injury is produced: nevertheless such reactions primarily promote localization and destruction of the invading agent, in addition to seducing those cells incriminated in the immunological response mechanism to the involved tissue site. An individual who manifests such damage is said to be hypersensitive to the antigen.

At present the terminology used to describe the various immunological states is both confusing and unsatisfactory. Fundamentally, the somatic system will exhibit a variable range of sensitivities towards invasion by foreign agents – from an insect bite to a full-blown bacterial infection – all of which manifest a particular sequence pattern: namely that the individual must first be sensitized (that is become allergic) by exposure to an antigen in the process of becoming immune, it taking a certain period of time, depending on the state, for antibody formation to take place. Following this, if the host is then further exposed to the same antigenic species the ensuing antigen–antibody interaction results in tissue injury. A system commonly used for classifying the types of sensitivities encountered depends on the time required to elicit the respective response – that is: (1) immediate hypersensitivity – here, as a result of the administered second dose, tissue damage occurs within minutes; (2) intermediate type – which is not a true immediate sensitivity state, since at least six to twelve hours elapse before a maximum reaction is manifested; (3) delayed hypersensitivity – where the reaction to the second antigen dose is slowly progressive, reaching its peak after 24–72 hours. Table 7.5 briefly summarizes the classification of the sensitivity states in the human species.

TABLE 7.5

Type	Example of state	Passive transfer agent	Causative immunoglobulin
Immediate (Prausnitz–Kuestner)	Asthma Urticaria Acute anaphylaxis	Serum	IgE
Intermediate	Protracted anaphylaxis	Serum	IgG or IgM
Delayed	Tuberculin	Viable lymph cells	Unknown

Immediate Sensitivity

Those antibodies precipitating an immediate response in the host's tissues are called Prausnitz-Kuestner antibodies or reagins, which belong to the IgE class of immunoglobulins. Their site of formation is not known with certainty. They are responsible for a variety of clinical syndromes ranging from hay fever, allergic asthma and anaphylaxis to sensitivities to drugs and foods.

These types of sensitivities can be divided for convenience into two groups depending on the route of entrance of the antigen. If the antigen is injected into the dermal layer of the host then the resultant response is termed cutaneous anaphylaxis. If, on the other hand, the antigen is introduced intravenously the response is referred to as generalized or systemic anaphylaxis, and this can lead to shock or asphyxia.

Cutaneous Anaphylaxis

This reaction is due to antigen–antibody interaction causing what is often called the weal and erythema response, since there is itching – probably due to the release of histamine, which causes a marked increase in capillary permeability – followed by the appearance of a weal surrounded by a zone of erythema; this occurs within minutes of the intradermal introduction of the antigen. Passive cutaneous anaphylaxis (PCA) can be demonstrated in the guinea-pig. Following the injection of purified antibody the animal is given an intravenous injection of dye (Evans' blue) prior to an intracutaneous injection of antigen. The response (PCA) is revealed by the accumulation of dye at the injection site – indicating an increase in permeability causing the passage of fluid containing the dye into the site with the formation of a circumscribed blue ring (Ovary, 1964).

Systemic Anaphylaxis

Here anaphylaxis is elicited principally by the release of mediator substances (see Table 7.6), of which five have been recognized: histamine, serotonin,

TABLE 7.6
Comparative Effects, and Mediators of Anaphylaxis in Man and Animals

Species	Symptoms of acute anaphylaxis	Pathology of acute anaphylaxis – principal organs involved	Principal mediator in anaphylaxis	Source of mediator and characteristics
Man	Tingling of scalp Flushing of skin Oedema of glottis Dyspnoea	Obstructive oedema of upper respiratory tract	Histamine SRS–A ? kinins	(1) *Histamine* Principal source: mast cells, platelets and basophils. Causes contraction of smooth muscle – can be blocked by antihistamines
Guinea-pig	Coughing, dyspnoea, convulsions and eventual collapse	Obstructed bronchioles, emphysema, hypotension and pancytopaenia	Histamines ? kinins	(2) *Serotonin* Principal source: mast cells, basophils, platelets and entero-chromaffin cells. Causes contraction of smooth muscle* – blocked by lysergic acid (LSD)
Mouse	Dyspnoea, hypothermia, paralysis of limbs and collapse	Lysis of intestinal entero-chromaffin cells. Decreased blood volume	Serotonin	(3) *Bradykinin* Precursor of α-globulin in plasma Contracts smooth muscle and rat uterus; is destroyed by peptides
Rabbit	Respiratory distress, convulsions and collapse	Pulmonary thrombosis, dilatation of right side of heart	Histamine SRS–A Serotonin	(4) *SRS–A* Is believed to be a lipid. Source unknown – thought to be released following breakdown of tissue Contracts guinea-pig ileum; no effect on rat uterus .Not inhibited by antihistamine
Rat	Respiratory distress, convulsions and collapse	Haemorrhage in ileum Low blood volume	Histamine Serotonin ? kinins	*contracts guinea-pig ileum (*see* Schulz–Dale Reaction)

267

bradykinin and lysyl-bradykinin, and a slowly reacting substance called anaphylaxis SRS–A. The mechanism by which antigen–antibody interaction motivates the release of these mediators is unknown, though it is conjectured that when an antigen reacts with cytophilic-based antibodies bound to cells such as the granulocytes or macrophages for example, local stimulation of the cell membrane occurs which precipitates the activation of cellular enzymes responsible for the release of the mediator substances. It is known that certain P-K antibodies are highly specific for mast cells and other cells of the granulocyte series which become fixed to the cell membrane for variable periods of time, and that – as a result of contact with a specific antigen – a mechanism is triggered which activates enzymes, releasing the mediators of inflammation in the immediate type of sensitivity reactions. The accumulation and persistence of large amounts of these mediator substances in the systemic circulation is prevented by the presence of mediator-destroying enzymes. Hence substances such as serotonin and histamine tend to act at local tissue liberation sites.

It has been demonstrated by passive anaphylaxis, using guinea-pigs as models, that the phenomenon is antibody dependent. Furthermore that a latent period is required for the antibodies to become fixed to specific cellular receptor sites at which subsequent antigen–antibody interaction will occur. The prime manifestation of anaphylaxis is the contraction of smooth muscle and vascular endothelial injury. Anaphylaxis may be summarized briefly as a state of shock, involving a particular organ or group of organs, resulting from a combination of immunopathological mechanisms mediated by a class or classes of immunoglobulins.

Anaphylaxis in man is often the result of drug therapy – that is, of the use of drugs, antisera, vaccines and antibiotics, to cite a few examples. Anaphylactic shock can be severe and even fatal. Symptoms commence with itching, flushing of the skin, headaches, respiratory distress, rapid fall in blood pressure and loss of consciousness which, if not checked, result in cardiac arrest. Sensitization may also occur naturally as a result of exposure to a certain agent present in the environment. For instance, some infants fed with cows' milk die from what is known as anaphylactic 'cot' death. It is thought that the feed is regurgitated and subsequently inhaled during sleep with the result that the child goes into a state of anaphylactic shock (Carpenter and Shaddick, 1965).

Probably the most widely reported and understood anaphylactic shock syndrome, however, is the one due to penicillin sensitization. Pencillin is known to be unstable, and sensitization is generally against the derivatives of penicillinic acid which are capable of conjugating with the protein of the host. The most common type of anaphylactic sensitization to penicillin is that of the Parusnitz-Kuestner type, though more dramatic and lethal circumstances can occur (Siegal, 1959; Fellner, 1968). Hence it is now customary for a clinician to take a history of allergy, and when such a likelihood is suspected, and a course of penicillin therapy indicated, a preliminary skin test is carried out to determine the sensitivity status of the recipient. Table 7.6 summarizes the possible mediators of anaphylaxis, and compares anaphylaxis in man with that in guinea-pig, rabbit, rat and mouse.

Schultz-Dale Reaction

As seen in Table 7.6, the principal shock mechanism in the anaphylactic response of most animals is the contraction of smooth muscle. The Schultz–Dale reaction can be used to demonstrate that an organ from a sensitized animal will respond specifically to the corresponding antigen *in vitro*. For example, the uterus excised from a guinea-pig 13 days after sensitization with an antibody and bathed in Ringer's solution will contract promptly when the appropriate antigen is added. This contraction is elicited, so it is assumed, by the release of histamine from the tissues, mast cells and platelets. Though mediators cause contraction of smooth muscle in guinea-pigs, in rats they cause relaxation.

Intermediate Sensitivity: The Arthus Reaction

In this category there are two main types of reaction: (1) the local Arthus reaction, first described by Arthus in 1903; and (2) the systemic Arthus reaction.

The Local Arthus Reaction

This can be induced in many animals including guinea-pigs, rabbits and mice, by administering a series of subcutaneous injections of foreign protein over a period of time. It is characterized by local constriction of the arterioles, retardation of the capillary blood flow, and aggregation of leucocytes and plate-lets to capillary vessel walls – resulting in injury and destruction of the capillary endothelium with ensuing haemorrhage and necrosis. It is believed that the formed antigen–antibody complex in the blood-vessel wall fixes $C'1$, and this has a chemotactic effect on polymorphonuclear leucocytes which subsequently surround and ingest the formed complexes: whether this is intravascular or extravascular, will depend largely on the molecular size of the reactants together with other considerations such as differential diffusion rates, tissue affinities, and clearance rates of the formed complexes. Thus, the result is the release of lysosomal enzymes from necrotic leucocytes which cause foci of necrosis in the local blood vessels, together with other inflammatory changes. IgG and IgM antibodies can produce this type of reaction. It would appear that the Arthus reaction may have evolved as a mechanism for eliminating immune complexes from the vascular system.

The Systemic Arthus Reaction

This is an immune situation occurring when large doses of antigen are administered subcutaneously to a hyperimmune recipient; this, in a matter of hours, results in circulatory failure and death. Autopsy findings commonly reveal a depleted concentration of serum C' together with increased fluid in the peritoneal cavity, circulatory stasis in the portal system, petechial haemorrhages in the lungs, serosae and various mucosae, and often massive haemorrhage into the gastro-intestinal tract. The antibody responsible, and the probable mechanism is evidently as for the local Arthus reaction.

Delayed-type Sensitivity

As the term implies the onset of the reaction is delayed, as compared to those responses first described. As yet the precise nature of the delayed type sensitivity and those antibodies responsible are unknown. In 1890 Koch working with the tubercle bacillus, observed that if these bacilli were inoculated subcutaneously into previously infected guinea-pigs a more intense inflammatory reaction occurred than in uninfected animals this now being known as the Koch phenomenon. Studies *in vivo* of delayed sensitivities have been made using the cutaneous and systemic tuberculin reaction.

The Cutaneous Tuberculin Reaction

If tuberculin (protein-purified derivative or PPD) is introduced into the dermal layer of a sensitized animal such as a guinea-pig, no visible reaction occurs until 6–12 hours have elapsed. Erythema and swelling at the site of inoculation then appear, becoming progressively larger and reaching maximal intensity after 24–72 hours. The reaction site will exhibit severe blanching, later followed by local haemorrhage and necrosis. Superficially this reaction resembles the Arthus reaction. However it must be remembered that in terms of reaction time the Arthus reaction is more rapid, usually appearing shortly after the administration of the antigenic dose – that is in some 2–4 hours – and subsiding after some 18–24 hours. Histologically the two reactions are significantly distinct. The Arthus reaction lesion is more oedematous and presents only a modest accumulation of inflammatory cells which are predominantly polymorphonuclear neutrophils. By contrast, the delayed-type lesion is less oedematous with a massive cellular infiltrate showing a higher percentage of macrophages and large lymphocytic cells in surrounding capillary blood vessels. Also the epidermis exhibits a hyperplastic picture. Antilymphocytic serum (ALS) is known to suppress the cutaneous tuberculin reaction in both the guinea-pig and man, but the precise nature of its action is unknown.

The Systemic Tuberculin Reaction

When a large dose of tuberculin (PPD) is injected intraperitoneally into a sensitized guinea-pig, a state of systemic tuberculin shock is precipitated after a delay of 3–4 hours, whereupon the animal exhibits prostration which, as the shock increases in intensity, is commonly followed by death. The state is characterized by lowered blood volume (increased haematocrit), hypothermia and circulatory failure. The exact nature of this shock state is still unknown. There is a demonstrable increase in vascular permeability which is considered to be the possible cause of shock, but the precise mechanism is still unknown. Histamine is apparently not the mediator of the reaction. ALS is seen to dampen the response—indicating the possible involvement of lymphocytes and macrophages in the immune mechanism.

Passive Transfer

Tuberculin sensitivity can be passively transferred from tuberculous guinea-pigs to normal allogenic guinea-pigs by injecting viable lymphoid cells into their peritoneum or blood stream. The recipient animal becomes sensitive as a result to both systemic and cutaneous tuberculin tests within 5–7 days of transfer. It has been observed that if the passive transfer cells are of peritoneal origin, the transfer is more successful than if spleen or lymph node cells are involved – all of which indicates the omnipotence of the macrophage which is found in high proportion in the former case. Though it must be stressed that passive transfer can be accomplished with lymphocytic cells also.

Contact Type Sensitivity

The acquisition of this type of delayed sensitivity is achieved via skin contact with sensitizing agents (allergens). A good example of this type is poison ivy sensitivity, caused by the natural product urushiol which gives rise to 'eczematous contact dermatitis'. The allergen of poison ivy appears to function as a hapten that conjugates with particular skin proteins resulting in epidermal hyperplasia, the formation of blisters and scaling of the epidermis. As yet the cell type responsible for the skin reaction has not been determined – though it has been observed that anti-lymphocytic serum (ALS) is capable of suppressing the ability of sensitized animals to mount a positive skin reaction.

Delayed-sensitivity Mechanisms

Delayed hypersensitivity can be differentiated from the immediate type by its inability to be transferred passively to normal recipients with serum from hosts with delayed sensitivity. However, transfer can be achieved by using lymphoid cells from sensitive donors. It must be admitted that a large obstacle facing immunologists endeavouring to understand delayed hypersensitivity concerns the difference between the findings in animals and man, and recognition of a definite consummation pathway.

With regard to the latter point — if more information can be found relating to the report of the existence of a transfer factor mediating in the delayed hypersensitivity mechanism then obviously a more satisfactory model could be assembled to aid subsequent research programmes.

The major difference between delayed sensitivity and the other types of sensitivities discussed is the factor of delay of onset. The reason for this lag period is multifactorial — and probably the most important single factor is that this time represents the period necessary for sensitive cells to respond to antigen. Present data indicates that both lymphocytes and macrophages contribute to the reaction, but exactly how and in which order is at present unfortunately unknown. Several workers have reported finding a transfer factor which they consider to be the active agent in the passive transfer of viable lymphoid cells. The nature of the factor is still rather uncertain, but it is known to be a substance with a molecular weight of less than 10 000, since it emerges as the initial peak on Sephadex G-25, dialyses through a cellophane membrane and is resistant to DNase and RNase. It is conjectured that this is a mediator factor — perhaps activating and recruiting normal recipient mononuclear cells, and in consequence responding to antigen by proliferating and possibly releasing tissue-damaging agents (Lawrence, 1960).

In Vitro Tests for Delayed Sensitivity Detection

If peritoneal exudate cells derived from tuberculin-sensitive guinea-pigs (containing a high percentage of macrophages) are placed in capillary tubes in small tissue culture chambers containing tuberculin — the cells from the tuberculin-sensitive animal are inhibited in their migration from the capillary tube, whereas migration of the cells from non-sensitive animals is unaffected (David *et. al.,* 1964). This and other *in vitro* studies seem to support the hypothesis that mononuclear cells from animals with delayed hypersensitivity are capable of transferring something to macrophages which, on contact with specific antigen, inhibits their migration.

Allergic Diseases

Serum Sickness

Serum sickness is a human self-inflicted disease, since it is caused by the therapeutic use of antigen — foreign animal serum. Historically this usage originated from the treatment of diphtheria with an injection of diphtheria antitoxin prepared from the horse (Von Pirquet and Schick, 1905). As a result of such an injection the recipient, some 10–12 days later, experiences a variety of symptoms including headache, urticaria, albuminuria, enlarged lymph nodes and swelling and aching joints. By virtue of the fact that the serum is often polyvalent – that is, it contains a variety of antigens and hence a number of different antigen–antibody systems (as indeed delayed types of sensitivity are involved) – the disease syndrome is of an undulating nature. This, since symptoms will tend to subside, then intensify, at intervals spread over a period of 14–28 days, as a consequence of the interaction of antibody as it is synthesized with the excess free antigen in the tissue fluids. Indeed sensitivities of Prausnitz–Kuestner and delayed types may be manifested. If individuals who have been exposed to antigen and previously manifested serum sickness are again given

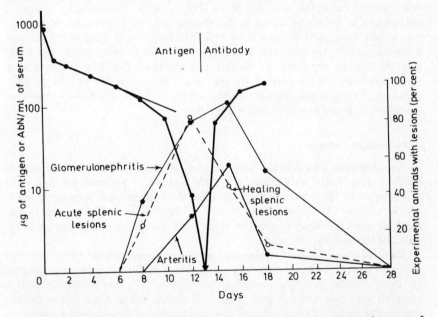

Figure 7.14. Relation of time of occurrence of tissue lesions to blood clearance of antigen and development of antibody. The curve showing the median blood clearance of antigen and the appearance of circulating antibody is represented by the thick solid line and refers to the ordinate on the left. Values after the thirteenth day represent antibody. The thin solid line projected from the antigen clearance curve at the sixth day shows the course the antigen would follow in the absence of an increased elimination of antigen by antibody. The incidence of splenic lesions at various times is represented by the broken line, and the incidence of arteritis and glomerulonephritis by the thin solid lines. These values refer to the ordinate at the right of the figure (from Germuth, 1953; reproduced by courtesy of The Rockefeller University Press)

the same antigen, this can precipitate a fatal systemic anaphylactic shock. For this reason a skin test – using the serum which is to be used for therapeutic purposes – is essential.

Germuth (1953) using experimental animals has shown the time course of serum sickness and the relationship of circulating antigen–antibody complexes to the development and healing of glomerular lesions etc. (*Figure 7.14*). Histologically the pathogenesis of glomerulonephritis of serum sickness appears to be due to the accumulation of these immune complexes in the epithelial aspect of the glomerular basement membrane.

Periarteritis Nodosa

Polyarteritis sometimes called periarteritis nodosa or nodular vasculitis is now generally accepted as a disease of allergic origin. The causative antigen is unknown. The lesions produced in this disease are very reminiscent of those of the Arthus type, which considered in conjunction with comparable induced lesions found in experimental animals would seem to substantiate the belief that the lesions represent an Arthus-type reaction. Features of the disease are disseminated inflammation of the small arteries giving rise to pyrexia, malaise and tachycardia. Possible complications of this inflammatory condition are nephritis, neuritis or myocardial infarction.

Auto-immune Disease

Auto-immune disease is the term used for an immune situation in which the host's system forms antibodies to a 'self' antigen – antigens which are the constituents of the tissues of the individual. An auto-immune response is one in which antibody is either demonstrable or capable of exhibiting an allergic reaction when challenged. On the other hand an auto-immune disease is the effect of the immune response.

Many of those diseases categorized as auto-immune appear to be due to a delayed sensitivity rather than to humoral antibodies. This statement is supported by the following principal experimental findings: (1) that passive transfer is achieved with cells and not with serum; (2) that antihistamine is ineffective as a therapeutic agent; (3) that delayed skin reactions to organ-specific antigens can in some cases be demonstrated; (4) that histopathologically the findings are comparable with those of delayed hypersensitivity reactions.

The present situation regarding the mechanisms involved in the causation of an auto-immune condition, and a stable classification of the diseases which are implicated is, for obvious reasons, not clear. One can say, at the risk of over-simplification, that there are considered to be two categories of auto-immune disease: (1) in which the antibody is formed against occult antigens – that is, a leak occurs of those antigens, which are considered not to be self-marked, thus motivating mobilization of antibody response mechanisms. An example of such a 'natural' auto-immune disease is sympathetic ophthalmia. Also such a disease may be experimentally induced using occult antigens – auto-allergic encephalomyelitis; (2) as the result of a situation in which antibody reacts with accessible antigen, which infers that some change has occurred in the response of antibody-forming tissues, or to the antigens themselves. Burnet (1959) in his clonal theory expresses the belief that the disease is the direct result of the mutation of forbidden clones capable of forming antibodies against the host's own antigens – that is, that the mutation has obliterated the memory of self-recognition.

In the example cited above of the natural auto-immune disease auto-allergic encephalitis – a disease of the central nervous system – the precise mechanism involved is not clear at present. The disease is characterized by scattered cellular infiltration and destruction of myelin in the brain and spinal cord. Since auto-allergic encephalitis and other diseases of a similar nature are often organ-specific it has been possible to induce the disease in experimental animals thereby affording workers an easily accessible model of the system. In the case of auto-allergic encephalitis the disease can be rapidly induced in guinea-pigs and rats by injecting emulsions of the CNS tissue involved mixed with Freund's adjuvant. The disease is manifest within two weeks, signalled by the development of CNS symptoms, such as paralysis and tremors. Histological findings are characterized by a perivascular infiltrate of cells, principally those of the lymphocytic series. Passive transfer can be accomplished with lymphoid cells. The antigens in the CNS appear to be the basic protein from myelin. For a full discussion of the immunological diseases see Bibliography and References (Samter, 1965).

Neoplasms and Abnormal Immunity Mechanisms

With the advent of inbred strains of animals, that is animals which are genetically identical, it has been established that a tumour can be transplanted from one animal to another in which it will flourish, whilst at the same time tumour-specific antigens are present (Prehn and Main, 1957). Prior to this the immune mechanism being observed was simply allograft immunity in which the tumour tissue transferred into a genetically different animal was simply being rejected like any other homograft. In fact if any genetic drift is allowed within these animals then this will lead to histo-incompatibility.

It is possible to detect tumour-specific antigens and immunize an animal against a tumour utilizing one of the following methods: (1) transplantation of a tumour which, after a certain period of time has transpired, can be surgically removed; (2) introduction of viable irradiated tumour cells; (3) introduction of a population of cells known to be insufficient to induce the formation of a tumour.

It has been shown also that tumours induced by a particular species of virus possess common tumour-specific antigens. For example, many of the various tumour types capable of being induced by the polyoma virus all possess common antigens specific for the polyoma virus. Hence one tumour can be used to immunize animals against other tumours originating from the same causative agent. By contrast those tumours induced by chemicals such as methylcholanthrene present the reverse of this – that is, cross-immunity is rare. Consequently such chemically-induced tumours clearly possess tumour-specific antigens – such that one tumour will effect immunization only against that particular tumour, and not against other tumours similarly induced.

MAMMALIAN BLOOD GROUPS

In 1900, Landsteiner as a result of his studies on mammalian blood, concluded that there were two distinct iso-antigens associated with human erythrocytes A and B – and that only one, both or neither may be resident at any given time on the red cell surface of the individual. Furthermore his results led him to note that whenever either or both of these iso-antigens is absent on the red cell of an individual, the specific iso-antibodies for the missing antigen are regularly present in the serum.

The blood groups named for the erythrocytic iso-antigens are A, B, O and AB. In group A cells there are two antigenic types, namely A_1 and A_2 – the anti-A antibody being composed of three antibody fractions called α, α_1 and α_2 respectively. The α antibody will react with both A_1 and A_2 cells; whereas α_1 antibody will react only with A_1 cells, and α_2 exclusively with A_2 cells. In fact A_2 accounts for some 20 per cent of the A population.

TABLE 7.7
Genotypes and Phenotypes Encountered in the ABO Blood Group System

Control gene	Frequency	Phenotype	Frequency	Erythrocytic antigen	Antibody circulating
AA	p^2	A		A Homozygous	Anti-B
AO	2pr	A	$p^2 + 2pr$	A Heterozygous	Anti-B
AB	2pq	AB	2pq	AB	–
BB	q^2	B		B Homozygous	Anti-A
BO	2qr	B	$q^2 + 2qr$	B Heterozygous	Anti-A
OO	r^2	O	r^2	–	Anti-A and Anti-B

There are three genes controlling the inheritance of the blood group substances; these are all alleles at one locus. These three allelic genes are A, B and O – A and B being dominant over O, which is recessive. Table 7.7 lists the genotypes and phenotypes encountered in the ABO blood group system.

It is believed that H is the basic substance from which the A and B antigens are derived. Group O will be seen to have the greatest amount of unconverted H substance – and it has been found that human erythrocytes of different groups exhibit decreasing reactivity with anti-H: thus, $O > A_2 > A_2B > B > A_1 > A_1B$, with the latter having the least amount (*Figure 7.15*).

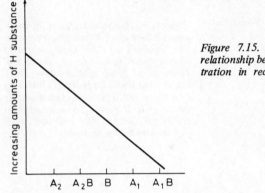

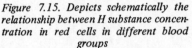

Figure 7.15. Depicts schematically the relationship between H substance concentration in red cells in different blood groups

The A, B and H substances are present as surface components of many epithelial and endothelial cell sites, and also in many somatic secretions such as saliva and sweat. About 80 per cent of the population are known as secretors – that is, their saliva contains A and B substances, or H substances in large quantities. Group O individuals, if they are secretors, will secrete H substance in their saliva and Group A, B and AB individuals, H plus A or B, or A and B substances respectively. There is a close relationship between Lewis (Le) blood groups and the secretor state; these antigens are believed to be determined by one of a pair of allelic genes, Le and le. The Le^a antigens are specified by the Le gene, and those individuals who have Le^a red cells are non-secretors, while those with Le^b [or Le (a– b+)] cells are secretors. It is considered that the blood group macromolecules of the ABO and Lewis system are assembled from a sugar mucopeptide on the superstructure of which the other various specific antigenic determinants of A, B, Le^a and H characters are welded by gene-controlled sequential enzymatic reactions (*Figure 7.16*).

The MNS blood group system. – The M and N antigens were first described by Landsteiner and Lewis in 1927, and two antibodies were recognized, anti-M and anti-N, which enabled a division of the population to be made into three genotypic groups, MM, MN and NN, with the corresponding phenotypes M, N and MN. Twenty years later Walsh and Montgomery described a factor called S which was closely associated with the MN system. This S antigen was found more

277

frequently on M and MN than on N cells. In the United Kingdom approximately 28 per cent of the population is group M, 22 per cent group N and 50 per cent group MN. Also 55 per cent of people have been found to be S-positive and the remainder S-negative.

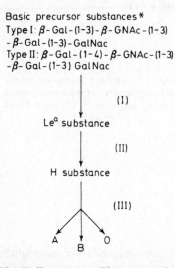

Basic precursor substances *
Type I: β- Gal - (1–3) - β - GNAc - (1–3)
- β- Gal - (1–3) - GalNac
Type II: β - Gal - (1–4) - β - GNAc - (1–3)
- β - Gal - (1–3) GalNac

(I)

Le^a substance

(II)

H substance

(III)

A   O
   B

Figure 7.16. Diagram depicting interrelationship between ABO and Lewis systems. The residues of the basic precursor carbohydrate chains in A, B, O and Lewis blood group substances are specified by the H and Le genes. (1) Indicates Lewis gene participation – incorporating fucose; (II) H gene acts to incorporate additional fucose; (III) indicates A, B co-dominant gene action

The Kell system. – The antigen first described by Coombs, Mourant and Race in 1946 was called Kell (K), but later an antibody anti-k (or anti-Cellano) was described which was capable of agglutinating K-negative cells. The Kell-Cellano antigens are avid antibody producers and will precipitate haemolytic transfusion reactions and haemolytic disease of the newborn.

TABLE 7.8
Genotypes and Phenotypes of the Duffy System

Genotype	Phenotype	Negro	Caucasian
		(percentage frequency)	
Fy^a Fy^a	Fy (a + b –)	9	17
Fy^a Fy^b	Fy (a + b +)	2	49
Fy^b Fy^b	Fy (a – b +)	22	34
Fy Fy	Fy (a – b –)	68	0

The Duffy system. – This system is known to have two allelomorphic antigens, Fy^a and Fy^b. Table 7.8 shows the phenotypes, genotypes and frequency percentages of the Duffy System in the Caucasian and Negro populations. Anti-Fy^a is frequently found in those individuals receiving multiple transfusions. This fairly common antibody occurs as the incomplete variety and is detectable with the antiglobulin test.

The Rh system was first described by Landsteiner and Weiner in 1940. The Rh factor is inherited as a simple dominance–recessive trait. Landsteiner and Weiner found that the red cells of the rhesus monkey (*Macaca mulatta*) injected into a rabbit produced an antisera capable of agglutinating the red cells of approximately 85 per cent of Europeans. Those people carrying the dominant gene are known as Rh positive, and the remaining 15 per cent as Rh negative. Fundamentally the inheritance of the Rh antigen is determined by three pairs of allelic genes CDE, cde of which three Rh genes are inherited from each of the respective parents. Recent findings have led to nomenclature difficulties and to the appearance of two theories purporting to explain the heredity of the Rh factors: these are Weiner's theory and the Fisher–Race theory. According to the Weiner theory, there are multiple alleles which each control the appearance of a particular antigen; whereas the Fisher–Race concept postulates the existence of closely linked genes on the same chromosome which is inherited as a unit. An endeavour is made in *Figure 7.17* to compare the two theories

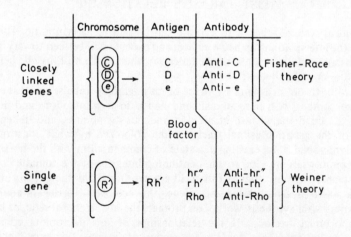

Figure 7.17. Comparison of the Weiner and Fisher–Race theories
in regard to the heredity of Rh factors

diagrammatically, and Table 7.9 compares the nomenclature of the two systems.

All those genotypes possessing gene D, and as a consequence the D antigen, which is by far the most potent, are said as individuals to be positive. In the absence of D, that is a genotype of cde/cde, the individual is Rh-negative (*see* Table 7.9 for alternative, yet rare, Rh-negative genotypes). The distribution of Rh antigens varies.

TABLE 7.9
Theories of the Heredity of Rh Factors Compared

| Fisher–Race | | Weiner | | |
Genes	Antigen	Gene	Antigen	Blood factor
CDE	c, D, E	R^Z	Rh_Z	Rho, rh$'$, rh$''$
cDE	c, D, E	R^2	Rh_2	Rho, rh$''$, hr$'$
CDe	C, D, e	R^1	R_1	Rho, rh$'$, hr$''$
cDe	c, D, e	R^0	Rh_0	Rho, hr$'$, hr$''$
CdE	C, d, E	r^y	rh_y	rh$'$, rh$''$
cdE	c, d, E	r''	rh$''$	rh$''$, hr$'$
Cde	C, d, e	r'	rh$'$	rh$'$, hr$''$
cde	c, d, e	r	rh	hr$'$, hr$''$

HOST–PARASITE RELATIONSHIP

In an ideal system, a host and parasite can be said to be symbiotic. Though in nature the interrelationship between host and parasite will be seen to vary greatly, in all instances the parasite is the beneficiary, whereas the host may be benefited, harmed or completely unaffected.

From birth onwards, and by virtue of his migratory impulses, man is exposed to close contact with new animals and hence to new infections and infection cycles. At birth the surface of his skin, mucous membranes, and the epithelial lining of the gastro-intestinal tract become colonized by what constitutes the normal microbial flora, existing in a state of commensalism with the host. In fact these commensals are vital to the continuing integrity of the somatic complex: for instance, vitamins such as folic acid are synthesized in the gut and absorbed by the active presence of bacteria. Hence this ecological barrier, in addition to performing vital synthesis work, also protects the host from subsequent invasion by opportunist bacteria. If the relationships of this microbiota community become unbalanced as a result of depression of the host's defence mechanism, such as when antibody production is depressed by x-ray irradiation, or the normal flora is altered by the action of chemotherapeutic agents — then micro-organisms constituting that flora can cause disease and be promoted to a pathogenic status (opportunists), in addition to invasion by known pathogenic species occurring.

Many pathogenic species of micro-organisms have adapted themselves to what has become their natural host which allows for propagation of the organism, without host destruction: that is, the host has acquired tolerance. This is seen to occur in the case of rabies virus which is widely disseminated amongst the vampire bats of North and South America — the classical natural hosts. It would now appear, however, that the perpetuation of the disease depends, in addition to the bats, on the existence of a reservoir of tolerant animals such as the Viverridae (mongoose), the Mustelidae (skunks) and other Chiroptera (bats). Investigations into this disease have revealed that under certain conditions the rabies virus can be transmitted via an aerosol route. In these tolerant reservoirs the nervous system does not appear to be attacked whereas the active disease state is the focal point of viral invasion resulting in paralysis, psychotic changes and agonizing death. Thus Man and the dog represent accidental hosts for rabies, for here the host's system is intolerant to the presence of the parasite, and as a result the host is unable to survive — achieving the opposite of what is really required by parasitism, since it is of no advantage to the rabies virus to kill the host.

Thus a potentially pathogenic micro-organism must be able to enter the host's system — the portal of entry depending on the type of pathogen. Entrance may be by penetrating the membranes of the respiratory tract, gastro-intestinal tract, or skin or by entering the blood-stream. The 'invasiveness' is a term used to describe the ability of an organism to penetrate the surface integument of the somatic system and thereby establish itself in the tissues of the host. Once penetration has been achieved, the organism must be capable of resisting the normal defence mechanisms of the host and must subsequently be able to multiply. Dissemination of an organism will depend largely on the ability of that organism to produce invasion-enhancing products such as toxins.

The results of an infectious disease or infestation are: (1) death; (2) recovery followed by immunity; or (3) recovery with tolerance of the parasite.

The Somatic Defence System

The Mucous Secretions

These secretions represent one of the first lines of defence, and contain significant concentrations of the antibacterial enzyme lysozyme, which is a basic protein of low molecular weight (some 15 000). The enzyme is also host-specific, and capable of attacking any bacteria, since the former has the ability to split a 1, 4 linkage between N-acetylmuramic acid and N-acetylglucosamine, and bacterial cell walls contain muramic acid (see Chapter 4). In Gram-negative bacteria the outer layer of the cell wall is predominantly lipoprotein — and the lysosomal action is only achieved through the initial attack on the outer coat by specific antibody and C'.

Phagocytes

These, like the mucous secretions, are of major importance, since they represent the first internal defence of the host in the form of the polymorpho-nuclear neutrophils, which are chemotactically seduced to the tissue injury sites. These cells exhibit a savagery probably unequalled in the microcosm — a capacity for ingesting and digesting material by a process known as phagocytosis. When a bacterial cell is engulfed, the formed phagosome receives hydrolyases from the lysosome which unites with the vacuole to form the phagolysosome, the digestive vacuole (*see Figure 4.7*, which depicts the fundamental scheme of the lysosomal apparatus).

The Reticulo-endothelial System

In a collection of specialized cells exhibiting phagocytic properties there are either circulating or fixed tissue phagocytes. The macrophage is a cell capable of differentiating as local needs dictate. Mobilization of the macrophage community is seen to be accelerated when the immune host is challenged a second time by an antigenic species.

Lymphocytic System

The lymph nodes and their ducts form a complex drainage system throughout the body, and are not infrequently a seat of infection and inflammation – causing nodal enlargement which can either be local or generalized.

Antibody Mechanisms

There are a number of antibody mechanisms which are able to effect: (1) neutralization of toxins; (2) production of specific antibodies exhibiting opsonic activity, which promote phagocytosis; (3) antimicrobial action of antibody together with complement participation, producing bacteriolysis.

Thus disease, such as infections, produce what can best be described as a disturbance in the equilibrium of normal somatic function. Though the precise mechanistic pathways of most diseases are not known, it would appear that they are largely dependent on the physiological location of the disease. In looking at disease it is important to consider some variable factors that will determine a host–parasite relationship, and thus influence the outcome of the infection. These factors can be classified as intrinsic and extrinsic.

The intrinsic factors may include the predisposition of an individual, family or race to a particular condition – or to be more specific the age of the particular host, his physiological state and immunological memory. With regard to the host's age, the neonate is reasonably resistant to infectious disease the status of which is controlled by specific humoral antibody, this being achieved by passive transfer of maternal antibody across the placenta or present in the colostrum. The host's management of an infectious disease becomes more effective with advancing age, until senescence – which may be described as the manifestation of an anamnestic mechanism. The immunological experience of the host is naturally a cumulative one, and is a function of age. The physiological state of the host will be governed by the extrinsic factors, which include such things as climatic and geographical location involving in turn a multitude of variables — such as, an inadequate water supply, malnutrition, vitamin deficiency, exposure to harmful radiation or toxic drugs — all of which depress the normal production of antibodies and phagocytes.

BIBLIOGRAPHY AND REFERENCES

Anderson, M. R. and Green, H. N. (1967). 'Tumour Host Relationships.' *Br. J. Cancer.* **21**, 27.

Asofsky, R. and Small, P. A. (1967). 'Colostral Immunoglobulin A: Synthesis *in vitro* of T-Chain by Rabbit Mammary Gland.' *Science, N.Y.,* **158**, 932.

Benkö, S., Lázár, G., Troján, I., Varga, L. and Petri, G. (1970). 'Prolongation of Skingraft Survival using Immune Serum.' *Nature, Lond.,* **226**, 451.

Billingham, R. E. (1963). 'Transplantation: Past, Present and Future.' *J. invest. Derm.,* **41**, 165.

— and Silvers, W. K. (1961). *Transplantation of Tissues and Cells.* Philadelphia; Wistar Institute Press.

— Brent, L. and Medawar, P. B. (1956). 'Quantitative Studies on Tissue Transplantation Immunity III: Actively Acquired Tolerance.' *Phil. Trans. R. Soc.,* **239**, 357.

Brent, L. (1967). 'Transplantation of Tissues and Organs.' *Brit. med. Bull.,* **21**, **97**, 182.

Bruton, O. C. (1952). 'Agammaglobulinaemias.' *Pediatrics, Springfield,* **9**, 722.

Burnet, F. M. (1959). *The Clonal Selection Theory of Acquired Immunity.* London; Cambridge University Press.

— (1961). 'Immunological Recognition of Self.' (Nobel Lecture — Dec. 12th 1960). *Science, N.Y.,* **133**, 307.

— (1967). 'Immunological Aspects of Malignant Disease.' *Lancet,* **1**, 1171.

— (1968). 'Evolution of the Immune Process in Vertebrates.' *Nature, Lond.,* **218**, 426.

Carpenter, R. G. and Shaddick, C. W. (1965). 'Role of Infection, Suffocation and Bottle Feeding in Cot Death.' *Brit. J. prev. soc. Med.,* **19**, 1.

Chase, M. W. (1945). 'The Cellular Transfer of Cutaneous Hypersensitivity to Tuberculin.' *Proc. Soc. exp. Biol. Med.,* **59**, 134.

— (1959). 'Models for Hypersensitivity Studies.' In *Cellular and Humoral Aspects of the Hypersensitive States.* New York; Harper.

Coombs, R. R. A., Mourant, A. E. and Race, R. R. (1946). 'A New Test for the Detection of Weak and 'Incomplete' Rh Agglutinins.' *Br. J. exp. Path.,* **26**, 255.

Coons, A. H., Leduc, E. H. and Connolly, J. M. (1955). 'Studies on Antibody Production I: A Method for the Histochemical Demonstration of Specific Antibody and its Application to a Study of the Immune Rabbit.' *J. exp. Med.,* **1102**, 49.

Cooper, N. R. and Fogel, B. J. (1967). 'Complement in Normal and Disease Processes.' *J. Pediat.,* **70,** 982.

Cruickshank, R. E. (Ed.) (1960). *Medical Microbiology.* London; Livingstone.

– and Weir, D. M. (Eds.) (1967). *Modern Trends in Immunology.* London; Butterworths.

David, J. R., Lawrence, H. S. and Thomas, L. (1964). 'Delayed Hypersensitivity *in vitro* II: Effect of Sensitive Cells on Normal Cells in Presence of Antigen.' *J. Immun.* **93,** 274.

– Al-Askari, S., Lawrence, H. S. and Thomas, L. (1964). 'Delayed Hypersensitivity *in vitro* I: The Specificity of Inhibition of Cell Migration by Antigens. *J. Immun.* **93,** 264.

Dreyer, W. J. and Bennett, J. C. (1965). 'The Molecular Basis of Antibody Formation.' *Proc. Natn Acad. Sci., U.S.A.,* **54,** 864.

Edelman, G. M. and Benaceraff, B. (1962). 'On Structural and Functional Relationships between Antibodies and Proteins of the Gamma-System.' *Proc. Natn Acad. Sci. U.S.A.,* **48,** 1035.

– and Gally, J. A. (1962). 'The Nature of Bence Jones Protein: Chemical Similarities to Polypeptide Chains of Myeloma Globulins and Normal Gamma-globulins.' *J. exp. Med.,* **116,** 207.

– – (1964). 'A Model for the 7S Antibody Molecule.' *Proc. Natn Aacd. Sci., U.S.A.,* **51,** 846.

Fahey, John L. (1965). *J. Am. med. Ass.,* **794,** 71.

Fellner, M. J. (1968). 'An Immunologic Study of Selected Penicillin Reactions Involving the Skin.' *Archs Derm.,* **97,** 503.

Fougereau, M. and Edelman, G. M. (1965). 'Corroboration of Recent Models of the IgG Immunoglobulin Molecule.' *J. exp. Med.,* **121,** 373.

Frisch, L. (Ed.) (1967). *Cold Spring Harb. Symp. quant. Biol.,* **32,** 1.

Germuth, F. G. (1953). *J. exp. Med.,* **97,** 257.

Gitlin, D. (1964). 'Protein Metabolism, Cell Formation and Immunity.' A Borden Award Address. *Pediatrics, Springfield,* **34,** 198.

– (1966). 'Current Aspects of the Structure and Function and Genetics of Immunoglobulins.' *A. Rev. Med.,* **17,** 1.

Gorer, P. A., Loutit, J. F. and Mickem, H. S. (1961). 'Proposed Revisions of 'Transplantese'.' *Nature, Lond.,* **189,** 1024.

Graber, P. and Miescher, P. (Eds.) (1962). *Mechanisms of Cell and Tissue Damage Produced by Immune Reaction.* New York; Grune and Stratton.

– and Williams, C. A. (1955). 'Methode Immuno-Electrophoretique D'Analse de Melanges de Substances Antigeniques.' *Biochim. biophys. Acta,* **17,** 67.

Häsek, M., Lengerová, A. and Vojtískocá, M. (Eds.) (1962). *Mechanisms of Immunological Tolerance.* Prague; Czechoslovak Academy Sciences.

Heidelberger, M. (1956). *Lectures in Immunochemistry.* London; Academic Press.

– and Kendall, F. E. (1935). 'Quantitation of Precipitin Reaction.' *J. exp. Med.,* **62,** 697.

Hilschmann, N. and Craig, L. C. (1965). 'Amino Acid Sequence Studies with Bence Jones Protein.' *Proc. natn Acad. Sci. U. S. A.,* **53,** 1043.

Himmelweit, F., Marquardt, Martha and Dale, Henry (Eds.) (1957). In *Collected Papers of Paul Ehrlich.* Oxford; Pergamon Press.

BIBLIOGRAPHY AND REFERENCES

Ishizaka, K. and Ishizaka, T. (1966). 'Physiochemical Properties of Reaginic Antibody I: Association of Reaginic Activity with Immunoglobulins other than IgA or IgG Globulins.' *J. Allergy.*, **37**, 169.

― ― and Hornbrook, M. M. (1966). 'Physiochemical Properties of Human Reaginic Antibody IV: Presence of a Unique Immunoglobulin as a Carrier of Reaginic Activity.' *J. Immun.*, **97**, 75.

Jerne, N. K. (1955). 'The Natural Selection Theory of Antibody Formation.' *Proc. Natn Acad. Sci. U.S.A.*, **41**, 849.

― Nordin, A. A. and Henry, C. (1963). In: *Cell Bound Antibodies. Conf. Nat. Acad. Sci. ― U.S. Nat. Res. Council.* Philadelphia; Wistar University Press.

Kabat, E. A. and Mayer, M. M. (1961). *Experimental Immunochemistry*, 2nd ed. Springfield, Ill.; Thomas.

Klein, G. (1966a). 'Tumour Antigens.' *A. Rev. Microbiol.*, **20**, 223.

― (1966b). 'Recent Trends in Tumour Immunology.' *Israel J. Med. Sci.*, **2**, 135.

Kunkel, H. G., Allen, J. C., Grey, H. M., Martensson, L. and Grubb, R. (1966). 'A Relationship between H Chain Groups of 7S Gamma Globulins and Normal 7S Gamma Globulins.' *J. exp. Med.*, **120**, 253.

Landsteiner, K. (1900). 'Zur Kenntnis der Antifermentativen Lytischen und Agglutinierenden Wirkungen des Blutserums und der Lymphe.' *Zentbl. Bakt. ParasitKde*, **27**, 357.

― and Weiner, A. S. (1940). 'An Agglutinable Factor in Human Blood Recognisable by Immune Sera for Rhesus Blood.' *Proc. Soc. exp. Biol. Med.*, **43**, 223.

Lawrence, H. S. (1955). 'The Transfer in Humans of Delayed Skin Sensitivity to Streptococcal M Substance and to Tuberculin with Disrupted Leucocytes.' *J. clin. Invest.*, **34**, 219.

― (Ed.) (1959). 'The Transfer of Hypersensitivity of the Delayed Type in Man.' In: *Cellular and Humoral Aspects of the Hypersensitive State.* New York; Hoeber.

― (1960). 'Some Biological and Immunological Properties of Transfer Factor.' In: *Cellular Aspects of Immunity.* Ed. by G. E. W. Wolstenholme and M. O'Connor. *Ciba Fdn Symp.* London; Churchill.

Milstein, C., Frangione, B. and Pink, J. R. L. (1969). 'Immunoglobulins.' *Nature, Lond.*, **221**, 145.

Moller, G. (1969). 'Regulation of Cellular Antibody Synthesis. Cellular 7S Production and Longevity of 7S Antigen-sensitive Cells in the Absence of Antibody Feedback.' *J. exp. Med.*, **127**, 291.

Mudd, S. (1970). *Infectious Agents and Host Reactions.* Philadelphia, Pa; Saunders.

Nisnoff, A., Wissler, F. C. and Lipman, L. N. (1960). 'Properties of Major Component of Peptic Digest of Rabbit Antibody.' *Science, N.Y.*, **132**, 1770.

Noelken, M. E., Nelson, C. A., Buckley, C. E. and Tanford, C. (1965). 'Gross Conformation of Rabbit 7S γ-immunoglobulin and its Papain-cleaved Fragments.' *J. biol. Chem.*, **240**, 218.

Ovary, Z. (1964). 'Passive Cutaneous Anaphylaxis.' In *Immunological Methods.* Ed. by J. F. Ackroyd. Oxford; Blackwell.

Pauling, L. (1940). 'A Theory of the Structure and Process of Formation of Antibodies.' *J. Am. chem. Soc.*, **62**, 2643.

Porter, R. R. (1959). 'The Hydrolysis of Rabbit Gamma Globulin and Antibodies with Crystalline Papain.' *Biochem. J.,* **73,** 119.

– (1962). 'Structure of Gamma Globulins and Antibodies.' In *Symposium on Basic Problems in Neoplastic Disease.* Ed. by A. Gellhorn and E. Hirschberg. New York; Columbia University Press.

Prehn, R. T. and Main, J. (1957). 'Immunity to Methylcholanthrene-Induced Sarcomas.' *J. natn Cancer Inst.,* **18,** 769.

Rhodes, J. M. and Sorkin, E. (1964). '*In Vitro* Studies on the Fate of Antigen. III. The *In Vitro* Fate of I^{131}-HSA in the Presence of Cells and Cell Extracts from Normal Immune and Tolerant Rabbits.' *Int. Archs Allergy appl. Immun.,* **24,** 342.

Rowe, D. S. and Fahey, J. L. (1965a). 'A New Class of Human Immunoglobulins. I. A Unique Myeloma Protein.' *J. exp. Med.,* **121,** 171.

– – (1965b). 'A New Class of Immunoglobulins. II. Normal Serum IgD.' *J. exp. Med.,* **121,** 185.

Samter, M. (1965). *Immunological Diseases.* Boston; Little, Brown and Co.

Scharff, M. D., Shapiro, A. L., Maizel, J. V. and Uhr, J. W. (1966). 'Polyribosomal Synthesis of the H and L Chains of Gamma Globulins.' *Proc. Natn Acad. Sci. U.S.A.,* **56,** 216.

Siegal, B. B. (1959). 'Hidden Contacts with Penicillin.' *Bull. Wld Hlth Org.,* **21,** 703.

Simonsen, M. (1957). 'The Impact on the Developing Embryo and New Born Animals of Adult Homologous Cells.' *Acta. path. microbiol. scand.,* **40,** 480.

Smith, R. T. (1968). 'Tumour-specific Mechanisms.' *New Engl. J. Med.,* **278,** 1207.

Steinberg, A. G. (1966). *Advances in Immunogenetics.* Ed. by T. J. Greenwalt. Philadelphia, Pa; Lippincott.

Tiselius, A. (1937a). 'A New Apparatus for Electrophoresis Analyses of Colloidal Mixtures.' *Trans. Faraday Soc.,* **33,** 524.

– (1937b). 'Electrophoresis of Serum Globulins: Electrophoretic Analysis of Normal and Immune Sera.' *Biochem. J.,* **31,** 1464.

– and Kabat, E. A. (1938). 'Electrophoresis in Immune Serum.' *Science, N.Y.,* **87,** 416.

Tomasi, T. B. and Zigelbaum, S. (1963). 'The Selective Occurrence of Gamma A Globulin in Certain Body Fluids.' *J. clin. Invest.,* **42,** 1552.

– Tan, E. M., Solomon, A. and Prendergast, R. A. (1965). 'Characteristics of an Immune System Common to Certain External Secretions.' *J. exp. Med.,* **121,** 101.

Turk, J. L. (1967). *Delayed Hypersensitivity.* Amsterdam; North-Holland Publishing Co.

Von Pirquet, C. and Schick, B. (1905). *Die Serum Krankheit.* Leipzig and Vienna; Deuticke.

– – (1951). *Serum Sickness.* (translated by B. Schick). Baltimore; Williams and Wilkins.

Wolstenholme, G. E. W. (1965). 'Complement.' *Ciba Fdn Symp.* London; Churchill.

World Health Organization (1964). 'Nomenclature for Human Immunoglobulins.' *Bull. Wld Hlth Org.,* **30,** 447.

Part V
Homoeostatic Function

8 – The Maintenance of Somatic Integrity

T. Hyde

(Introduction – Enzymes – Hormones – Regulation of carbohydrate metabolism – Integration of glucose homoeostasis – Regulation of lipid metabolism – Regulation of protein and nucleic acid synthesis – Tissue growth and repair – Water, electrolyte and acid–base balance)

INTRODUCTION

It is clear that, for a multicellular organism to survive at all, each individual cell must be capable of monitoring and maintaining its own structure and metabolism; but this cannot occur entirely in isolation, and many of its functions must be related to those of numerous other cells. The situation is therefore quite different from that obtaining for a free-living cell. This need to exercise control over both the milieu *intérieur* and *extérieur* requires the cell to be both apparently static in nature, in order to maintain continuity of form, and yet also dynamic, in order to be responsive to the varying pressure of its own environment and those of the organism as a whole.

The various vital reactions must proceed at an orderly and balanced rate, and the supply of structural and energy-rich compounds must be related strictly to demand. The preservation of this state of balanced dynamism is called homoeostasis, and in the intact organism it operates at several different levels: between whole organs, between the individual cells of a tissue, and within the cell itself. The adjustments to be make by the cell are therefore rather complex, especially as many of the various external and internal factors must, to a considerable extent, be integrated within it.

Obviously, in order that the alterations of metabolism be appropriate both in kind and quantity, information must be made available concerning the present state of the cell metabolism. Feedback mechanisms, very often of the negative type, are therefore a fundamental feature of homoeostasis, and are present in many forms. The requisite information is transmitted either by molecules of specific structure (and this mode may occur within or between cells) or by the nervous system.

In order to maintain the dynamic structures of the cell metabolism in a rapidly responsive state, it is often a rather characteristic feature of biological control mechanisms for the information to affect two opposed systems, so that the final level of the particular process controlled is really determined by the sum of the action of exciting and inhibiting factors. The most obvious and striking examples of this occur in the central nervous system in connection with voluntary control, but the interplay of such factors in chemical systems will be seen later. Again, in multicellular organisms, it is common to find that a system affording rapid fine control is often superimposed on one which enables major adjustments to be made relatively slowly.

In the following discussion, a general resumé of the various control mechanisms is presented first, as a background to the more detailed account of particular metabolic systems.

ENZYMES

All metabolism is dependent on the presence of enzymes within the cell to provide the requisite rapid interconversions of substances at biological temperatures and pH.

The type and rate of conversion depend on the particular enzyme and clearly the enzymes must operate in a defined sequence. In part, this may be achieved by an enzyme having absolute specificity for its substrate; or perhaps by the separation of the enzymes by endoplasmic reticulum membranes. The sequential placing of the cytochrome system of enzymes on the cristae of the mitochondria, the sequestration of nucleases in lysosomes and the presence of glucose-6-phosphatase in liver microsomes are examples of this.

Although enzyme kinetics in general will not be discussed here, it is relevant to review some particular aspects (*see also* Chapter 2). It is clear that the rate of formation of the product of an enzymatic reaction can be affected in one of three ways:

(1) by alterations in the substrate or product concentrations;
(2) by changes in enzymatic activity;
(3) by increased synthesis of the enzyme.

Alteration in the Substrate or Product Concentrations

In the case of (1), it will be recalled that the law of mass action applies to enzyme reactions, and thus an increase in substrate will increase the rate of formation of the product; conversely an accumulation of product will slow and eventually reverse the reaction — even if the enzyme activity is constant. This follows the reactions of enzyme with substrate:

$$E + S \rightleftharpoons ES \rightleftharpoons P + E$$

Changes in Enzymatic Activity

The activity of an enzyme, estimated from the Michaelis dissociation constant (K_m), or from the maximal velocity (V_{max}) can in normal functions be altered in the several different ways now described.

Competitive inhibition. — Such inhibition occurs when the competitor occupies the specific binding site on the enzyme which is normally occupied by the substrate molecule. The action is a simple blockade. It does not affect the configuration of the enzyme molecule, and is reversible by mass action. The competitor substance has a similar steric configuration to the substrate. A physiological example is that of succinic acid dehydrogenase, for which malonate competes with succinic acid, thus reducing the rate of conversion of the latter.

Non-competitive inhibition. — This is unrelated to substrate concentration and does not appear to be important in control mechanisms (*Figure 2.6*) except in the form of allosteric inhibition.

Allosteric inhibition. — Monod and Jacob gave this term to a particular type of inhibition occurring in regulatory enzymes. Here the product of a series of reactions affects only one enzyme; this may in fact occur at a point early in the metabolic sequence. A simple example is the hydrolysis in rat liver of 3-phosphoserine by serine phosphatase — the product serine specifically inhibits the enzyme in a non-competitive fashion:

Allosteric or regulatory inhibitors are not steric analogues, nor do they bind to the active site, but instead exert their influence on some other specific portion of the enzyme molecule. They may alter the K_m, the number of binding sites or — occasionally — the V_{max}, and it is thought that these modifiers may induce some configurational change in the molecule, similar to the changes known to occur in the haemoglobin molecule on oxygenation. Enzymes exhibiting this phenomenon are always regulatory in type, and they can not only be inhibited but also activated by modifiers.

Some regulatory enzymes show very marked changes in structure under the influence of certain modifiers, so that distinct active and inactive forms with different characteristics have been recognized.

Isoenzymes. – It is known that some enzymes may exist in forms which show physical differences in such characteristics as heat stability, electrophoretic mobility and kinetic constants. For the enzyme lactic acid dehydrogenase at least, these differences seem to have a physiological significance. The less electrophoretically mobile forms of this enzyme are found in liver and skeletal muscle, and these can readily and preferentially form lactate from pyruvate which is produced in large amounts from the glycolytic pathway. The more mobile forms, occurring in heart muscle, are inhibited by quite low pyruvate concentrations, and the pyruvate produced must be largely oxidized in the tricarboxylic acid cycle.

Thus the enzymes appear to be adapted to function under low and high oxygen concentrations respectively. However, little is known of the physiological and genetic factors which influence the ratio of the various isoenzymes in a tissue, and it may well be that small alterations in the ratio in a tissue are important in the adjustment of metabolism (Wolstenholme and Knight, 1969).

Increased Synthesis of the Enzyme

Enzymes are proteins and are synthesized in the ribosomes. Increased production may therefore follow: (1) an increase in the production of mRNA by the nucleus, probably with formation of new ribosomes; (2) the direct stimulation of ribosomal activity.

Induction. – This phenomenon, induction, was accounted for in the hypothesis of Monod and Jacob. In bacterial cells, the regulatory gene is directly affected by simple metabolites, and often these may be substrates, although this is by no means invariable. In *E. coli,* methyl-β-galactoside will induce the formation of β-galactosidase, and this occurs within several minutes. It should be noted that although several different inducers may affect one particular regulatory gene, only the same enzymes are induced by their action. Furthermore, induction cannot occur at all unless some enzyme is already being synthesized in the cell: in other words the capability for enzyme formation must be present already, and induction is impossible if the regulatory or structural genes are blocked and inoperative.

In mammalian cells induction and repression may in part be mediated in such a way, but it is known that various hormones can influence regulating genes, and this is a mechanism of equivalent importance.

Increase in ribosomal activity. – This may occur directly by the influence of cell metabolites, but also seems to be a function of hormonal stimulation.

Regulatory Enzymes

The flow of metabolic transformations is mediated by many enzymes, but it is now quite clear that only a limited number of enzymes are involved in the actual control of the rate at which metabolism proceeds. These are known as regulatory, or rate-limiting, enzymes; they may be regarded as functioning rather like valves in a pipeline, and they have certain features in common.

Modifiers. – They may undergo either inhibition or activation by modifiers, and this is of the allosteric type already described. The modifiers may be products of the rate-limiting reaction, or derivatives from later reactions in the sequence. However, many instances occur in which the products of linked reactions, such as the formation of ATP, can also be implicated.

Multiple active sites. – They are characteristically composed of a number of sub-units, each of which has catalytic activity, and they therefore contain more than one active site per molecule.

Unusual kinetic features. – Their kinetics are unusual, in that instead of the reaction velocity showing a hyperbolic relationship to substrate concentration, they usually exhibit a sigmoid curve. They may also require a threshold concentration of substrate to be present before any conversion occurs at all, and they show greater sensitivity to changes in substrate concentration below the K_m.

The shape of the [V]/[S] curve is a direct result of the presence of a number of binding sites, the affinity of which varies with the number of sites already bound. A similar relationship is seen in the oxygenation of haemoglobin, where several binding sites are available for oxygen.

Largely irreversible reactions. – The reactions which they govern are generally irreversible, primarily because they are exergonic, with a large loss of free energy occurring.

Characteristic 'committed step'. – The regulatory enzymes tend to be sited at the first point in a linked metabolic sequence. This is sometimes called the 'committed step' because subsequent reactions are simply intermediate exchanges, following automatically from the first reaction without any branching points.

Cross-over of mass action ratios. – They may be identified by the 'cross-over' of mass action ratios. The mass action ratio = [product]/[substrate]; as the reaction velocity increases with substrate concentration it can be seen that the mass action ratio is inversely proportional to the reaction velocity, and that when the latter is zero (at equilibrium), the mass action ratio is maximal and equal to the equilibrium constant of the enzyme.

When the regulatory enzyme activity is low, the other enzymes will be working below capacity, and their mass action ratios will be high and close to the equilibrium constant. However the converse is true of the regulatory enzyme, as it will be working at its full — although limited — capacity, and its mass action ratio will therefore be low, due to substrate accumulation. When the activity of the regulatory enzyme is increased, the rate of flow is increased, and the mass action ratio therefore rises towards the equilbrium constant.

However, because of the acceleration, the product concentration of the preceding enzyme falls, and the substrate concentration of the succeeding enzyme rises, so that their mass action ratio falls. Thus the change in mass action ratio of the regulatory enzyme is the reverse of that of the other enzymes, and this change is further distinguished by being much greater in magnitude.

Transport Enzymes

In connection with the regulatory steps, mention must be made of transport systems. These are probably complexes of carrier proteins and enzymes situated in cell membranes, both at the surface and in mitochondrial and endoplasmic structures. They are concerned with the transfer of compounds such as glucose and amino acids, or of ions such as sodium. The transfer is made across a concentration or energy gradient, and thus requires a source of energy, usually ATP. The system functions overall as an enzyme, and generally conforms to standard Michaelis–Menten kinetics.

Regulatory function in cell metabolism. – However, a number of transport systems have a regulatory function in cell metabolism. One of major importance is the transfer of glucose to the interior of the cell, and in muscle and adipose tissue this is activated by the presence of insulin. An example within the cell is the citrate transport mechanism in the mitochondrial membrane which controls the transfer of citrate from mitochondria to cytoplasm, where it has an inhibitory effect on phosphofructokinase. This system is activated by malate.

Clearly, the point at which all regulatory factors and information operate is on the conversion rate of certain pace-making enzymes; to borrow Sherrington's phrase these represent the 'final common path' for regulation. At the cellular level, therefore, three main types of control may be distinguished.

(1) Regulation of entry of substances into the cell, by transport systems. The mode of regulation of the transport mechanisms is unknown, but hormones such as insulin and steroids are involved.

(2) Direct repression or activation of a regulatory enzyme by a metabolite. This is a common form of control which allows rapid fine adjustment of enzyme reaction rates.

(3) Repression or induction of enzyme synthesis. This is effected by a metabolite from within the cell, or a metabolite or hormone from without,

influencing the regulatory gene. Although this is a rapid process in bacteria, it is characteristically much slower in the mammalian cell, and allows of coarse long-term control.

HORMONES

Hormones are chemical transmitters which, when released into the extracellular fluid, pass to other cells ('target cells') and produce specific effects on their metabolism. They are an extremely important link in the homoeostatic control of many body functions, and the means by which overall chemical control of the cellular metabolism is achieved.

They may be disseminated in the blood stream (for example, thyroxine), or may diffuse and act quite locally (as in the case of acetylcholine); some, such as the anti-diuretic hormone, have a specific target tissue, while others, such as cortisol, may act on many types of cells. Ultimately many of their effects are exerted on enzyme activity, through their action in the cell either: (1) on transcription; (2) on translation; or (3) directly on the enzyme or enzyme complex; some indeed may operate at all three levels. These actions are considered in more detail elsewhere. A few hormones act through the agency of intermediary hormones. Some hormones may also affect the polysomal production of structural proteins by action at nuclear or cytoplasmic levels.

Hormones may be produced in separate glandular structures, and the products of the pituitary, adrenal and thyroid glands and of the gonads are examples, while others may arise from small groups of tissue (for example, insulin), or from diffusely distributed cell systems (as in the case of secretin and prostaglandins). The neural and pituitary control of the major hormones is discussed in the pages which follow: the secretion of some hormones, however, seems to be directly related to their metabolic effects, without the intervention of complicated feedback links. The relationship between calcium and parathyroid hormone is a good example of this. Thus hormones are secreted by their parent tissues in response to metabolic demands; the controlling factor may be only one metabolite, but usually numerous factors, both chemical and neural, are concerned.

Pituitary and Hypothalamus

The direct control of hormonal output by a metabolite provides relatively little flexibility in response, especially to external factors, and such flexibility must be an important consideration where a hormone has multiple effects. External factors, and many metabolic components, are monitored by the nervous system, and the brain may adjust hormone levels: (1) directly, as in the release of adrenaline from the adrenal medulla under sympathetic stimulation; or (2) indirectly, by way of the hypothalamic—pituitary axis.

The pituitary gland develops from an ectodermal diverticulum in the roof of the pharynx (Rathke's pouch) and a downgrowth from the diencephalon, the former producing the anterior lobe, the latter, the posterior lobe. The *posterior lobe* has direct neural connections with the hypothalamus; it stores and releases anti-diuretic hormone and oxytocin. The *anterior lobe* produces a number of hormones, the more important being growth hormone, adrenocorticotrophic hormone (ACTH), thyrotrophic hormone (TSH), and the gonadotrophic hormones. Growth hormone apparently has direct actions on basic metabolism in the cell (page 304) but the function of the others is to control the synthesis of hormones in the adrenal, thyroid and gonads respectively.

There is now much evidence that these hormones are partly under the influence of the hypothalamus, which releases specific chemical transmitters. These pass to the anterior lobe of the pituitary by way of the portal vessels in the stalk of the gland, where they cause the release of the particular trophic hormone. Specific, but differing, areas in the hypothalamus and pre-optic nuclei selectively take up cortisol, oestrogens, and thyroxine, and it is thought that these are receptor sites for the feedback mechanism.

This mechanism has been studied in most detail for cortisol. If the adrenal output of cortisol falls, sensitive cells in the median eminence of the hypothalamus are stimulated to release a corticotrophin-release factor (CRF). This is a polypeptide closely related to vasopressin; in man, lysine-vasopressin has CRF activity. This compound passes along the portal vessels to the anterior lobe, stimulating the formation and release of ACTH, which in turn increases the adrenal synthesis of cortisol, thus maintaining the normal output. However, the plasma level of cortisol is not constant but shows a diurnal variation; this phenomenon appears to be of hypothalamic origin and is related to the sleep period, the peak value for plasma cortisol being reached some 6 hours after waking. Adjustment of this rhythm following alterations in the sleep pattern does occur, but takes 4–6 weeks to be complete. Administration of cortico-steroids for therapeutic purposes results in a suppression of ACTH release, and this may last for some days or weeks after discontinuing treatment.

This system is of physiological importance for basal homoeostasis, but it has been shown that any stressful situation – for example, one brought about by surgery – acts directly on the pituitary and overcomes the normal feedback mechanism; further, ablation of the hypothalamus does not markedly reduce the plasma 17-hydroxysteroid level so that a considerable degree of control must be exerted at pituitary level, independently of the hypothalamus. Also, there is some evidence for a separate inhibitory area located in the hypothalamus.

Although thyroxine probably does have some direct inhibitory action on the release of TSH by the pituitary, there is clear evidence for the presence of a TSH-releasing factor, acting in a similar way to CRF and polypeptide in nature.

FSH and LH are also under hypothalamic control; in the male some unknown testicular component is probably responsible for the feedback of information for FSH release, while in the female the controlling factor appears to be oestrogen. In both sexes FSH levels rise to high values following castration. Specific areas in the hypothalamus have been shown to take up oestrogens, and there is experimental evidence that releasing factors are secreted by hypothalamic cells and pass through the portal vessels to the pituitary. In the female there is a rhythmic production of FSH and LH, producing the oestrus cycle, whereas in the male it is constant. This rhythm is also of hypothalamic origin; the female pattern is present in both male and female animals before birth, but conversion to the male pattern follows exposure to androgens, and is irreversible.

The hormones secreted by the anterior pituitary have actions on specific target organs, and their action in releasing a final, metabolically active, hormone from the cells of these tissues appears to be mediated by cyclic AMP (cyclic 3', 5'-adenosine monophosphate), the so-called 'second messenger'. Certainly, not only TSH, ACTH, LH and vasopressin, but other hormones such as adrenaline and parathormone, will cause a rise in the concentration of cyclic AMP in their respective target cells. Furthermore, experimental administration of cyclic AMP, and even of its analogues, can reproduce some of the effects of the hormones. In muscle, adrenaline operates directly through this mechanism, and the cyclic AMP which is produced activates phosphorylase.

However, for the trophic hormones, cyclic AMP appears to act as an intermediary messenger within the cell, so that the presence of the trophic hormone stimulates cyclic AMP formation (possibly by some action on adenyl cyclase or its synthesis) the increased levels of which in turn cause the release of the particular hormone. It is possible that this mechanism allows the effects of the hormone to be modified by the availability of ATP and thus of energy production within the cell. It is interesting that prostaglandins also exert effects on adenyl cyclase (and hence on cyclic AMP formation), but these hormones oppose the stimulatory action of adrenaline, TSH, ACTH and other hormones (Editorial, 1968).

It can be seen from the foregoing facts, that the homoeostatic loops involved in the control of hormone secretion are open to modifying influences at many points, thus providing considerable flexibility of response.

THE REGULATION OF CARBOHYDRATE METABOLISM

Glucose, as glucose-6-phosphate, is a key compound in the metabolism of the living cell, partially because its catabolism provides ATP, the major source of energy, and partially because it is the source of many important structural precursors and compounds such as ribose and lipids. It is clear that the precise control both of the flow of free glucose to the cell, and of its concentration and utilization in the cytoplasm, is of vital importance in the organism. The discussion of this control will begin with the main pathways of intracellular metabolism, after which the general body homoeostasis of carbohydrate will be treated. The relationship with lipids, and the changes in the diabetic state, are considered later.

Intracellular Regulation of Glucose Metabolism

Glycolysis

The Embden-Meyerhof pathway is the major route for the conversion of glucose to pyruvate; this is shown in *Figure 8.1*. Only the regulatory enzymes are shown; for other details a biochemical text should be consulted. It will be

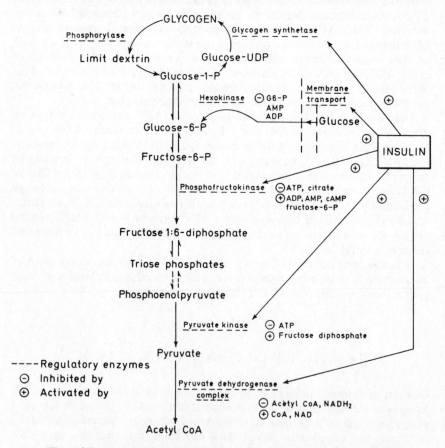

Figure 8.1. General scheme of glycolysis: only regulatory enzymes are shown

298

remembered that only the liver and kidney contain glucose-6-phosphatase, and can therefore secrete free glucose.

On reference to *Figure 8.1* it will be seen that glucose uptake is controlled by the cellular membrane transport process, and by the subsequent phosphorylation of glucose to glucose-6-phosphate by hexokinase. The next controlling step is the conversion of fructose-6-phosphate to fructose-1:6-diphosphate. This reaction is exergonic and virtually irreversible; it is mediated by phosphofructokinase, an enzyme which is inhibited by ATP and citrate, and activated by ADP, AMP, cyclic AMP and by its substrate, fructose-6-phosphate.

The next regulatory point is the conversion of phosphoenolpyruvate to pyruvate, by pyruvate kinase. This enzyme has an important regulatory function in yeast, but its position in the mammal is less clear, although there is some evidence that it can be inhibited by fatty acids and activated by insulin, and may be concerned in the regulation of the supply of $NADPH_2$.

The final stage in the pathway is the formation of acetyl CoA from pyruvate by the action of the pyruvate dehydrogenase complex. This complex functions as a regulatory enzyme and it is activated by coenzyme A and NAD_2 and inhibited by acetyl CoA and $NADH_2$.

Glycogenesis and Glycogenolysis

These reactions in muscle are summarized in *Figure 8.2*, omitting the branching and debranching systems. They represent the conversion of glucose-1-phosphate to a branched, less osmotically active, storage form, glycogen, and the reformation of glucose-1-phosphate from glycogen by a different pathway. Glucose-1-phosphate is attached to uridyl diphosphate, and by repetitive transfer a chain of glucosyl 1:4 units is built up under the influence of the regulatory enzyme glycogen synthetase (UDPG-glycogen transglucosylase). This enzyme exists in two forms, the very active I form, unresponsive to glucose-6-phosphate, and the D form, which is dependent on glucose-6-phosphate for activation, and is therefore linked to intracellular metabolism, unlike the I form. Conversion of the I to the D form is by the action of cyclic AMP, which is itself produced by adenyl cyclase, activated by adrenaline. The reverse change is brought about by another enzyme, glycogen synthetase phosphatase, which is activated by insulin, and inhibited by glycogen.

The breakdown of glycogen is regulated by phosphorylase (which acts on the 1:4 links of the side-chains) and which also exists in two forms, a and b. The a form is an enzymatically active tetramer, which is unaffected by intracellular signals such as AMP, ATP and glucose-6-phosphate, and the b form, which is a dimer and requires AMP for activation. The conversion of form b to a is controlled by activated phosphorylase kinase, which is itself activated by cyclic AMP. The formation of cyclic AMP is stimulated by adrenaline, which increases adenyl cyclase activity. The phosphorylated form of phosphorylase kinase requires the

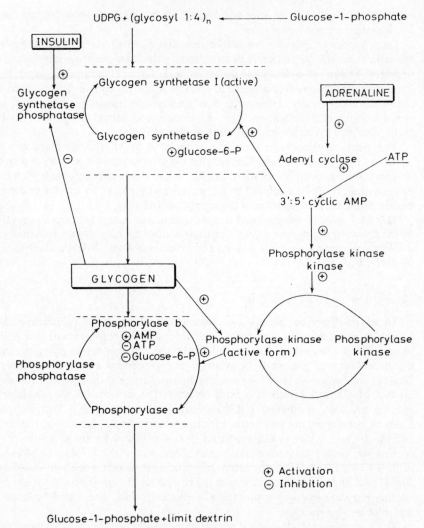

Figure 8.2. Regulation of glycogen cycle in muscle

presence of glycogen for its full activity. Phosphorylase systems in the liver are essentially similar, except that glucagon is the major stimulator of phosphorylase conversion.

The systems outlined above, despite their complexity, function as a single integrated unit; the actions of hexokinase, glycogen synthetase D, phosphorylase b, and phosphofructokinase being linked by the concentrations of glucose-6-phosphate and fructose-6-phosphate. Glucose-6-phosphate inhibits hexokinase and phosphorylase b, and activates glycogen synthetase D; inhibition of hexokinase by glucose-6-phosphate raises the free intracellular glucose, but this rise in substrate concentration does not remove the inhibitory effect: the glucose-6-phosphate level must be reduced by conversion to glycogen or fructose-6-phosphate before further free glucose is phosphorylated.

A rise in fructose-6-phosphate concentration activates phosphofructokinase, but this may be inhibited by ATP and citrate, which are products of later stages in glucose catabolism. Obviously a fall in ATP concentration will activate phosphofructokinase (because of accumulation of ADP and AMP) and the concentration of both fructose-6-phosphate and glucose-6-phosphate, will fall. This causes activation of hexokinase and phosphorylase b, and inhibition of glycogen synthetase D. Thus phosphorylation of glucose and utilization of glucose-6-phosphate will be increased. Insulin, amongst other actions, promotes glucose uptake by facilitating its transport, by converting glycogen synthetase D to the active I form, and possibly by suppression of pyruvate carboxylase and phosphofructokinase.

The adjustment of glycogen metabolism in muscle is connected also with the availability of ATP. Glycogenesis is active when ATP levels are high; in active muscle the concentration of ATP falls; this, together with a concomitant rise in AMP and fall in glucose-6-phosphate, activates phosphorylase b, with consequent formation of glucose-6-phosphate. However, adrenaline is also secreted during exercise, and this converts phosphorylase b to the more active a form, so that glycogen phosphorolysis is proceeding at its maximum rate. At the same time the glucose-6-phosphate level remains low (because of increased utilization) and inhibits the activity of glycogen synthetase D, while the influence of adrenaline via cyclic AMP also causes conversion of the active I form of glycogen synthetase to the much less active D structure. The reverse sequence of events occurs on cessation of muscle activity. As one might expect, the control of glycogenolysis in the liver differs from that in muscle in being more closely related to substrate concentrations than to the phosphorylase activity.

From the foregoing, it will be appreciated that glycogen functions as a buffer in glucose homoeostasis, this role being quantitatively more important in muscle than in liver. At least six well-defined diseases of glycogen metabolism have been described in which there is complete absence of one particular enzyme (Symposium, 1968).

Gluconeogenesis

Although some enzymatic steps in glycolysis are virtually irreversible because of the energy changes involved, it is nevertheless possible for the pathway to be reversed by means of other enzymes, so that the unfavourable reactions are by-passed. A source of energy is necessary for this reversal, and is provided by ATP. The overall change in the glycolytic process is as follows:

$$\text{glucose} + 2ADP + 2P \longrightarrow 2\,\text{lactate} + 2ATP$$

becomes:

$$2\,\text{lactate} + 6ATP \longrightarrow \text{glucose} + 6ADP + 6P$$

From reference to *Figure 8.3* it will be seen that four enzymes are involved in the conversion of pyruvate to glucose: pyruvate carboxylase, phosphoenolpyruvate carboxylase, fructose-1:6-diphosphatase and glucose-6-phosphatase.

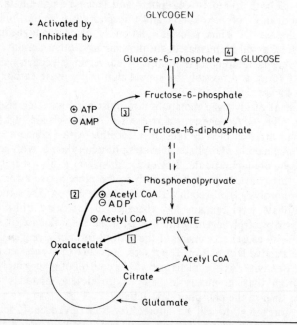

Regulatory enzyme	Repressed by:	Induced by:
(1) Pyruvate carboxylase	Insulin, ADP	Glucocorticoids, adrenaline, glucagon
(2) Phosphoenolpyruvate carboxylase	Insulin	Glucocorticoids
(3) Fructose-1:6-diphosphatase	Insulin, substrate	Glucocorticoids, adrenaline, glucagon
(4) Glucose-6-phosphatase	Insulin	Glucocorticoids

Figure 8.3. Simplified scheme of gluconeogenesis, showing the regulatory enzymes

302

These are all regulatory enzymes, their action is essentially irreversible and, as noted later, they work as a functional unit, the activity of all being stimulated indirectly by glucocorticoids. Liver is the only organ containing a significant quantity of fructose-1:6-diphosphatase, and is therefore the major site of gluconeogenesis.

Hexose Shunt (Phosphogluconate Pathway)

This system is considered in detail in biochemical texts; briefly it provides for the complete oxidation of glucose, the overall changes being: 6 glucose-6-phosphate + 12 NADP + 7H$_2$O $\rightarrow$ 6 CO$_2$ + 12 NADPH$_2$ + 5 glucose-6-phosphate + H$_3$PO$_4$

The cycle is probably important more as a source of NADPH$_2$ and ribose, than in energy production. All the reactions are freely reversible, except for the decarboxylation of 3-keto-6-phosphogluconate. This step may not have a regulatory function as the shunt system occurs mainly in tissues in which lipid synthesis is active (for example, adipose tissue and adrenal cortex), and it is probable that the presence of active lipogenesis, with its requirement for NADPH$_2$, stimulates active degradation via the pathway, so that a major controlling factor is probably the availability of NADP.

The Effects of Hormones on Carbohydrate Metabolism

Before passing to a general review of carbohydrate homoeostasis, some pertinent details of the action of the several hormones involved must first be given.

Insulin

The effects of insulin on the regulation of carbohydrate metabolism are still not known in detail; it acts chiefly on muscle, adipose tissue and liver. In both muscle and liver its major action is in stimulating the uptake of glucose by the cells. In muscle and adipose tissue it does this not by activation of phosphorylation, but by increasing the capacity of the transport system, by some direct effect not mediated by RNA. Insulin is known to bind easily to cell surfaces, and it is thought that it may have some allosteric effect on the enzymes in the cell membrane.

It should be remembered that insulin has no effect on glucose transport in liver or brain, but it does activate a specific glucokinase in the liver which increases hepatic uptake. Some direct activation (other than induction) of hexokinase, glycogen synthetase and possibly other regulatory enzymes may also be due to insulin (*Figure 8.3*).

It was suggested earlier that the regulatory enzymes acted as though functionally linked. In the liver it is known that insulin activates hexokinase, phosphofructokinase and pyruvate kinase. This action can be blocked by actinomycin D and is thus probably due to the induction of synthesis of these enzymes, possibly by a direct effect of insulin on an operon, although this has not been proved. Insulin also suppresses the synthesis of enzymes involved in gluconeogenesis and so tends to increase uptake and intracellular utilization of glucose. Apart from this action on the transcription process, it may also have effects at the translational level in the polysomes.

Adrenaline and Glucagon

These hormones have basically similar actions on carbohydrate metabolism. They both cause a rapid rise in blood sugar, by direct activation of adenyl cyclase and thus, via cyclic AMP, of phosphorylase. Glucagon acts in this way only in the liver, whereas adrenaline is effective in both liver and muscle cells. The action of adrenaline is very rapid, activating phosphorylase in the isolated heart within 45 seconds. There is also some evidence which suggests that glucagon stimulates gluconeogenesis in the liver.

It is now known that glucagon can be a potent stimulator of insulin secretion *in vivo* (Samols, Marri and Marks, 1965); however, there is some evidence which throws doubt on its possible physiological role as a stimulator of insulin release, and it has been suggested that only glucagon of intestinal origin has this effect *in vivo*.

Glucocorticoids

These compounds have effects on carbohydrate metabolism which are not clearly understood, but a major effect appears to be the stimulation of gluconeogenesis. The catabolic effect of steroids on proteins provides the basic material. In liver, at least, these hormones appear to induce a related group of enzymes involved in gluconeogenesis (*Figure 8.3*), possibly by de-repression of an operon involving their synthesis. Glycogen synthetase activity is increased, and there is some evidence that they can inhibit the metabolism of pyruvate.

Growth Hormone

It was formerly thought that growth hormone was only of importance in children, but since the advent of immunoassay methods capable of detecting circulating levels, it has been realized that it is normally present in adult blood, and that fluctuations in the level occur over short time intervals. Except for their anabolic action on protein metabolism, the actions of insulin and growth hormone on fat and carbohydrate metabolism are opposed, as growth hormone inhibits glycolysis in muscle and stimulates gluconeogenesis in the liver.

The changes in the blood following insulin administration are shown in *Figure 8.4;* it is possible that the reciprocal changes in growth hormone and insulin levels are of some importance in the fine control of glucose flux (Editorial, 1967a). Certainly the release of growth hormone is stimulated by hypoglycaemia, amino acids, stress and ACTH, and it has been suggested that its release is more a result of stress than a part of normal physiology (Zahnd, Nadeau and von Muhlendahl, 1969).

The uncertainties as to the role of the various hormones are mainly due to difficulties in analysis, and these difficulties will only be resolved by the introduction of more specific methods.

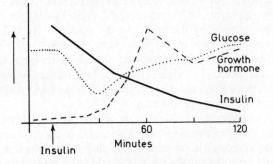

Figure 8.4. Growth hormone and glucose responses following the injection of insulin in a normal human subject (after Greenwood and Landon)

INTEGRATION OF GLUCOSE HOMOEOSTASIS

Although hormones and other influences operate ultimately within the cell, the action of the intracellular mechanisms can be seen more clearly when they are regarded from the standpoint of the integrated function of organ systems. In carbohydrate metabolism, a primary objective is the need to regulate glucose flux within fairly narrow limits, and the attainment of this is reflected in the stability of blood glucose levels. Four hormones at least (glucagon, adrenaline, growth hormone and glucocorticoids) tend to raise the blood sugar, adrenaline and glucocorticoids being more active under stress. Their action is opposed only by insulin, and it is possible that this preponderance of hyperglycaemic agents reflects the vulnerability of the brain to any reduction in its glucose supply.

The various intracellular and hormonal mechanisms discussed are closely linked to the level of glucose in the blood. In the early fasting state, glucose levels fall to normal values, with a concomitant drop in insulin levels; the level of growth hormone probably rises, but recent work has shown that glucagon output is not increased, as was formerly thought (Vance, Buchanan and Williams, 1968). If fasting continues, adrenaline is released, and glucocorticoids are secreted by the adrenal cortex, under the action of ACTH — released (as part of the stress reaction) by the pituitary. The result of these changes is that the peripheral utilization of glucose is diminished in many tissues (but not in the brain), and the liver becomes a net producer of glucose; both gluconeogenesis and glyco-genolysis are increased, the latter partially by the direct action of adrenaline (*Figure 8.2*), the former initially by the fall in insulin levels, and later by the action of glucocorticoids (*Figure 8.3*). If the starvation continues, these changes are reinforced, glycogen stores become depleted, and more precursors for gluconeogenesis (glycerol, fatty acids, lactate, amino acids) are mobilized from peripheral tissues, mainly by the action of the glucocorticoids. Thus the liver is the major source of glucose in the basal state. However, after several days of fasting, most tissues become adapted to utilizing substrates other than glucose for energy production. Brain, red cells and renal medulla utilise glucose preferentially, but other substrates can be used during fasting: in brain these may account for up to 40 per cent of the energy production.

When food is ingested, blood glucose levels rise, and insulin secretion is stimulated. The above changes are then reversed, and the liver uptake of glucose increases, while glycogen is re-formed and gluconeogenesis diminishes. In the peripheral tissues both the uptake of glucose and glucose utilization are increased under the influence of insulin.

REGULATION OF LIPID METABOLISM

In this section the homoeostatic regulation of triglyceride metabolism is discussed, together with its relationship to carbohydrate metabolism.

Neutral fats, or triglycerides, are esters of fatty acids with glycerol, of the general structure:

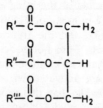

where R represents a fatty acid side-chain. Fatty acids are ultimately derived from carbohydrate or from dietary sources, and glycerol is indirectly formed from glucose. The interrelationships of lipid and carbohydrate are summarized in *Figure 8.5*.

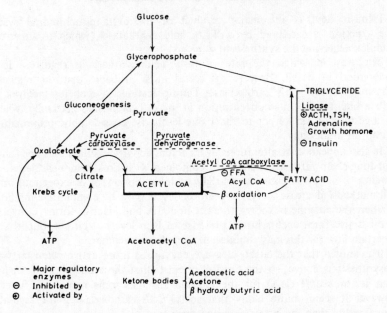

Figure 8.5. Schematic relationships of carbohydrate and triglyceride metabolism: only major regulatory enzymes are shown

Fatty Acids

The *de novo* synthesis of fatty acid occurs chiefly in the liver, gastro-intestinal tract, lung and bone marrow. The fatty acids are derived from carbohydrate by way of the intermediate compound acetyl CoA. The synthesis proceeds in two main steps. Much simplified, these are:

(1)   acetyl CoA + ATP → malonyl CoA + ADP + phosphate

This step is under the control of a biotin-dependent enzyme, acetyl CoA carboxylase. Although acetyl CoA is produced from pyruvate in the mitochondria, fatty acid synthesis is chiefly extra-mitochondrial. The precise mechanism by which acetyl CoA is transported is not known, but two main mechanisms have been suggested:

(*a*) acetyl CoA is converted in the mitochondria to citrate, which is transported across the mitochondrial membrane, and is then reconverted to acetyl Co A and oxalacetate by the citrate-cleavage enzyme.

(*b*) acetyl CoA combines with carnitine, forming acetyl-carnitine which traverses the membrane; after this the complex dissociates.

(2)   malonyl CoA + acetyl CoA → palmitic acid

307

Palmitic acid is the major product of the extra-mitochondrial system. The reaction is catalysed by a 'fatty acid synthetase', which is an enzyme complex achieving the synthesis in some six stages.

The way in which the rate of fatty acid synthesis is regulated is not understood in detail. However, it seems most probable that the regulatory enzyme is acetyl CoA carboxylase. During fasting, in diabetes mellitus, and with a high fat diet, its concentration in the cell diminishes markedly, whereas on ingestion of carbohydrate the tissue levels rise considerably, presumably as a result of induction.

It also appears that alterations in its actual activity occur; it has been shown that long-chain fatty acids alone, or combined with CoA (whether they are the product of *de novo* synthesis, of triglyceride catabolism, or of exogenous origin) will markedly decrease acetyl CoA carboxylase activity by competitive inhibition. Further, the enzyme is known to exist in active and inactive forms; conversion to the active form occurs in the presence of high levels of citrate, but it is still uncertain how far this may function in a regulatory manner.

It is known that the citrate-cleavage enzyme is more active when fatty acid biosynthesis is active, so the possibility exists that the availability of substrate (that is, of acetyl CoA) for the enzyme also has some regulatory function. However it seems more likely that acetyl CoA carboxylase is the regulatory factor, rather than the citrate transport system.

The catabolism of fatty acids occurs chiefly by the process of β-oxidation, a cyclic five-step process in which two molecules of acetyl CoA and five of ATP are produced at each turn of the cycle. This occurs exclusively in the mitochondria; the fatty acids are transported across the mitochondrial membrane in combination with carnitine, and this may be a regulatory point.

Triglycerides

Triglycerides are synthesized in large amounts in adipose tissue. They are also synthesized in the intestine and liver, from which they are released into the circulation as chylomicrons or as low-density lipoproteins respectively. The plasma is cleared of these by the action of lipoprotein lipase, which is activated by heparin, and which hydrolyses the triglycerides to free fatty acids (FFA) and glycerol, in which form they are absorbed by the tissues. In one form of hyperlipidaemia in man, this enzyme is absent and the plasma remains milky, with high triglyceride levels. It is possible that lipoprotein lipase may have some regulatory action *in vivo*.

Most of the circulating low-density triglyceride is taken up by adipose tissue, which contains lipoprotein lipase. However, it also contains a hormone-sensitive lipase which acts, not on circulating lipid, but on neutral fat already stored within the cell. This lipase is activated by adrenaline, noradrenaline, ACTH, TSH, and growth hormone, and it is inactivated by insulin. The action of these hormones on lipase activity in adipose tissue is probably mediated in the cells by cyclic AMP.

In the fasting state, the triglycerides stored in adipose tissue are mobilized, and it appears that the hormone-sensitive lipase is the regulatory enzyme of major importance in controlling this process. During fasting, insulin levels are low, and the action of growth hormone, adrenaline and other hormones stimulates the lipase, with the result that there is an increased rate of hydrolysis in the adipose tissue cell. Normally much FFA released is re-converted to triglyceride before it can leave the cell, but this depends on the availability of carbohydrate (as glycerophosphate), and in fasting the supply is curtailed. The result is that glycerol diffuses out into the plasma, because adipose tissue contains very little glycerokinase, and therefore it cannot re-utilize glycerol directly to form triglyceride. FFA also leaves the cell, but becomes bound to albumin due to its water insolubility. The glycerol is metabolized in the liver; some of the FFA is taken up by the peripheral tissues and utilized as fuel, while the liver takes up the remainder, and either converts it to low-density lipoprotein, or catabolizes it to ketone bodies. The latter process is quantitatively more important in fasting, and the liver is the only tissue containing the necessary enzymes for ketone formation. The rate of tissue uptake is determined by the amount of FFA bound to albumin, and thus ultimately by the rate of lipolysis. The reverse situation, of course, obtains in the fed state, with diminished fat mobilization and increased triglyceride synthesis.

Activation of the sympathetic nervous system causes a rapid release of FFA, apparently due to the release of noradrenaline by sympathetic nerve fibres many of which are present in adipose tissue. Other fat-mobilizing substances have been described, such as prostaglandins which are ubiquitous in body tissues and of which E_1 prostaglandin is known to induce lipolysis in man, while fractions have been isolated from human pituitaries which have lipolytic activity (Burns, Hales and Hartree, 1966), and similar substances appear in the urine during fasting.

Obesity. – Obesity may briefly be mentioned here. It is now a common and important disease with a high morbidity, and while it is usually due to an excessive caloric intake, there are some people who undoubtedly handle their metabolic fuels more efficiently and in whom even starvation results in a smaller weight loss than is the case with normal people. The factors involved are still unknown, and for this reason the discovery of a strain of genetically obese and hyperglycaemic mice has aroused considerable interest as there appears to be loss of some regulatory function. These mice have multiple defects, but their adipose tissue shows a much greater than normal increase in fatty acid biosynthesis even when fasting. Fat mobilization is reduced, and this may be due to the low lipase and much increased glycerokinase content of the adipose tissue cells, with a consequent tendency for any FFA to be converted to triglyceride within the fat cell rather than liberated. Some such imbalance may be a factor in humans, but many other factors must be involved: for example it has been suggested that the regulation of caloric intake may be related to the size and number of adipose tissue cells (Hirsh and Gallian, 1968).

Lipid and Carbohydrate Metabolism and Diabetes Mellitus

It will be evident by now that the homoeostatic functions of carbohydrate and lipid metabolism are closely related and integrated (*Figure 8.5*).

For example, when glucose flux is high, fat mobilization and fatty acid catabolism are suppressed. This occurs because;

(1) glycerophosphate is available in the adipose tissue cell for the synthesis of triglyceride from FFA;

(2) insulin secretion is stimulated, and thus the lipolytic action of hormone-sensitive lipase is diminished with the result that little FFA is released for utilization by the peripheral tissues or liver. The reduced amount of FFA available to the liver not only restricts its output of low-density lipoprotein but also diminishes the generation of ketone bodies from acetyl CoA;

(3) FFA in tissues other than liver and fat is also diverted from the pathway of β-oxidation into triglyceride and complex lipid synthesis.

Fatty acid biosynthesis is also stimulated by increased carbohydrate catabolism, probably again because of the greater availability of glycerophosphate: this removes FFA and fatty acyl CoA, both of which strongly suppress acetyl CoA carboxylase, and so the conversion of acetyl CoA to malonyl CoA is increased.

Thus the net overall effect of carbohydrate intake is increased lipogenesis, with a decrease in lipolysis and ketone body formation.

Reversal of these changes occurs in the fasting state, with the result that (except for brain, renal medulla and erythrocytes) the tissues no longer utilize carbohydrate as an energy source, but instead use acetyl CoA derived from fatty acids and ketone bodies. Because of this switch-over in energy-producing substrates, it is possible for the mammal to survive prolonged fasting, with just sufficient insulin being secreted to balance lipolysis and formation of ketone bodies against utilization, and to allow the essential basal glucose flux to proceed.

On the other hand, FFA has multiple effects on carbohydrate metabolism: in liver and kidney at least FFA inhibits hexokinase, phosphofructokinase and pyruvate dehydrogenase. The increased acetyl CoA derived from the FFA inhibits pyruvate dehydrogenase, and stimulates pyruvate carboxylase activity, and thus promotes gluconeogenesis. Many other reactions occur, but in general terms the net effect of FFA is to block carbohydrate metabolism; however, FFA does not appear to have any such action in brain.

310

Diabetes Mellitus

This is a disease in which there is a relative or absolute deficiency of insulin. Insulin levels are low in fasting, as already noted, but in the classical juvenile form of diabetes, there is a complete, or almost complete, lack of insulin. The result of this can now be appreciated: overall glucose uptake and catabolism by the cell is reduced (*Figure 8.1*) and lipolysis is stimulated. As glycerophosphate is not available within the cell, a state of imbalance results in which large amounts of FFA are liberated into the plasma. FFA is extracted by the liver, but because of the reduction in glycolysis, the liver is unable to esterify all the FFA reaching it and thus diverts it into the formation of ketone bodies, producing the characteristic ketosis and acidosis. However, some FFA is esterified to triglyceride, and is released by the liver causing the typical lipaemia: in uncontrolled diabetes the plasma may appear like cream.

Because of the diminished membrane transport, glucose uptake by the cell is diminished, and glucose levels in the blood rise considerably, especially as gluconeogenesis is increased and the liver is unable to catabolize glucose to any extent, or to store it as glycogen. The glucose in the plasma exceeds the renal threshold, thus inducing a solute diuresis, with the typical polyuria and glycosuria after which the disease is named.

REGULATION OF PROTEIN AND NUCLEIC ACID SYNTHESIS

The overall metabolism of proteins is dealt with in biochemical texts, while the more general aspects of the regulation of cell growth are discussed on page 171. Here we are concerned solely with some aspects of the intracellular regulation of protein and nucleic acid synthesis.

The mechanism of protein synthesis is described in detail on p. 205, but briefly, the information contained in an operon on the DNA strand in the nucleus is transcribed into mRNA, this transcription being controlled by an operator gene, which can in turn be activated or inhibited by a regulator gene, probably acting through a repressor protein. The mRNA passes to the cytoplasm as a ribosome complex, and becomes attached to endoplasmic membrane, where it takes up amino acids (bound to tRNA), and its encoded information is translated into the polypeptide sequence of specific proteins. These are mainly structural proteins or enzymes, and on their continued production the integrity of the cell depends. The synthetic machinery may be regulated at some point in the nucleus, or else within the cytoplasm: that is, at the transcriptional or translational level, and it may be affected by: (1) products; (2) substrate; or (3) hormones.

Products

It is well recognized that the product of an enzyme sequence may regulate synthesis of a regulatory enzyme, and this appears to occur not only in bacterial but in mammalian cells also. A good example of this process in mammalian cells is given by δ-amino laevulinic acid synthetase (ALA synthetase). This enzyme is the first step in the synthesis of porphyrins and thus of haem, and it has been shown to play a major regulatory role in the hepatic synthesis of porphyrins. The accumulation of haem inhibits the formation of ALA synthetase, and thus the synthesis of porphyrins and of the regulatory enzyme is affected by the presence of the end product (Granick, 1966). This is due to haem acting as a co-repressor, and activating the repressor protein in some way so that formation of more mRNA for ALA synthetase by its structural gene is shut off.

Substrates

The induction of galactosidase was described earlier as an example of substrate induction in a bacterial cell. This may occur in mammalian cells, but the presence of substrates may also have effects on the rate of RNA and protein catabolism. This has been clearly shown to occur in the liver, by Munro and others: they found that the RNA content of liver was increased when fasted animals were fed protein, and they were able to demonstrate that this was due to a reduced rate of RNA breakdown. It was also accompanied by an increase in the amount of endoplasmic reticulum.

Hormones

Much protein synthesis, especially of regulatory enzymes, is under hormonal control. Some hormones, such as oestrogen and thyroxine, act only on specific tissues, but others such as insulin affect many. Some features of their mode of action suggest that the translation of the genetic code in mammalian cells differs in many ways from that in bacterial systems. The work of Tata (1969) and others has provided some understanding of the way in which hormones influence protein synthesis.

It has been known for some time that hormones will stimulate protein synthesis at the ribosomal level *in vivo*, but will not do so in cell-free systems *in vitro*. Hormones also stimulate RNA synthesis, of course, and some of their effects could be ascribed to selective regulation of mRNA synthesis. However, other processes seem to occur. For example, oestrogen or thyroxine will stimulate the nuclear RNA polymerase and thus the synthesis of all classes of nRNA; rRNA cistrons appear to be the most sensitive, but some mRNA is also formed. Much of the nRNA does not leave the nucleus, but appears to be degraded within it; however, under hormonal stimulation, an increased amount of rRNA and specific mRNA is transferred to the cytoplasm. A marked increase in the number of ribosomes occurs, and there appears to be a concomitant increase in microsomal membranes, and an increase in their synthetic activity.

The newly-induced ribosomes are more firmly bound to the endoplasmic reticulum membranes, and the different types of ribosomes induced by different hormones appear to remain segregated.

Hormones generally have little effect at the translational level, except on those polysomes specifically induced by the particular hormone, and in these polysomes they do appear to have a direct activating effect. Thus hormones appear to increase the rate of formation of the whole synthetic machinery, although the new ribosomes generated produce only a specific range of proteins, depending on the inducing hormone. The way in which hormones function in the nucleus is unknown, but they do not directly affect the transcription process in isolated nuclei, and it is probable that the structural genes are affected only in some indirect manner.

DNA Synthesis

The control of RNA synthesis is largely bound up with protein synthesis, but the regulation of DNA synthesis is a distinct and complex problem. To some extent it is related to mitosis, but it is clear that DNA synthesis, mitosis and cell growth must be partially under independent constraints. The factors regulating their interdependence are still poorly defined, but DNA synthesis is, of course, a necessary preliminary to cell division (although not necessarily a triggering factor *per se*). Some of the regulatory mechanisms in DNA synthesis are known and will be briefly described here.

The first stage in the synthesis is the formation of the purine and pyrimidine nucleoside monophosphates: these are preferentially synthesized from small molecules such as aspartate rather than from the parent bases. These ribonucleosides form the corresponding deoxyribonucleotides which are then phosphorylated to the triphosphate compounds; under the influence of a DNA polymerase and with DNA template, these nucleotides are polymerized into a DNA strand. Control appears to be exercised at the stage of nucleotide synthesis, rather than polymerization. For example, although the normal liver contains almost no deoxyribonucleotide, it does contain some DNA polymerase. Much of our knowledge of DNA synthesis in mammals comes from studies of rat liver regenerating after partial hepatectomy (page 316).

The first regulated step in the synthesis of the pyrimidine nucleosides is the conversion of carbamyl phosphate and aspartate to ureidosuccinate by the enzyme aspartate carbamyl transferase; this enzyme is directly inhibited by cytidine triphosphate, a later product (*Figure 8.6*). Similarly, the first regulatory enzyme in purine nucleoside biosynthesis appears to be amidophosphoribosyl transferase, which catalyses one of several steps in the formation of inosine monophosphate from ribose, ATP, glutamine and glycine. This enzyme also exhibits inhibition by product, as its activity is markedly decreased by adenosine monophosphate and guanosine monophosphate.

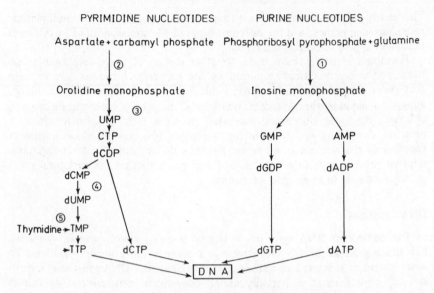

Figure 8.6. Diagram of DNA synthesis and regulation

In regenerating liver the content of aspartate carbamyl transferase begins to increase very shortly after partial hepatectomy, and is one of the earliest changes known to occur. Orotidine monophosphate (*Figure 8.6*) is formed from ureidosuccinate, and is converted to uridine monophosphate (UMP) by the enzyme orotidine monophosphate decarboxylase; the activity of this enzyme is also inhibited by its product, UMP.

The conversion of the nucleoside diphosphates of cytosine, adenine and guanine to the deoxyribose forms is also under negative feedback control, by the inhibitory action of the deoxytriphosphate derivatives themselves; similarly, thymidine triphosphate (TTP) has an inhibitory effect on the formation of dexoycytidine diphosphate from cytidine diphosphate, these being precursors of TTP (*Figure 8.6*). Thymidine is a pyrimidine base unique to DNA, and its conversion to thymidine monophosphate by the enzyme thymidine kinase is also inhibited by its triphosphate derivative (TTP). Finally, thymidine monophosphate may be formed from deoxycytidine diphosphate:

$$dCDP \longrightarrow dCMP \xrightarrow{\text{deoxycytidylate deaminase}} dUMP \longrightarrow TMP$$

The activity of the enzyme deoxycytidylate deaminase is also inhibited by TTP; in regenerating liver it appears some 12 hours after partial hepatectomy, but does not reach its peak activity until 48 hours have elapsed.

It is of interest that the various enzymes involved in nucleotide synthesis tend to show a sequential induction and suppression in regenerating liver, with the activity of first-stage enzymes such as aspartate carbamyl transferase appearing first, and the various kinases (which phosphorylate the monophosphate nucleosides to the triphosphate forms) appearing later. The decline in enzyme activities follows a similar course. The mechanism of this induction and suppression is quite unknown at present.

The inhibitory feedback loops thus appear to be geared to utilization of the nucleotides. The presence of all four nucleotides is a necessary preliminary for DNA polymerization, and as long as incorporation of the nucleotides continues, the low concentration of intermediate compounds will prevent inhibition of their formation. Obviously, once DNA synthesis ceases, the supply of intermediates will be suppressed in turn, both by direct inhibition of enzyme activity by the products, and also by represssion of enzyme synthesis.

TISSUE GROWTH AND REPAIR

Although the present discussion of homoeostasis necessarily centres on the basic and better known parts of metabolism and function, some other aspects must be mentioned briefly here, if only to indicate the limitations of our knowledge. In biological evolution it was evidently a major step for single cells to become united in functional multicellular groups; most importantly, from our present purview, it meant that (to speak teleogically) the individual cell had to 'allow' some of its metabolic processes to become subservient to, or organized towards, the 'common good'.

The individual cells must still be allowed to grow, yet the rate of growth must be controlled, and their general morphological relationships must reflect, and be adapted to, the overall function of the animal. Even in a simple radial form such as hydra, mechanisms must exist which allow the individual cells to grow together and to differentiate, and yet which arrest and control growth when the animal has reached its optimal size. The factors involved probably differ from those which determine the growth and division of individual cells. This 'problem of growth' is of central interest and importance in biology today; for an understanding of this a solution must be sought to a number of closely related phenomena, which will be very briefly discussed in this section.

Embryonal Growth

The co-ordinated growth and differentiation of the embryo is a remarkable phenomenon, and since the work of Spemann and Mangold it has become increasingly clear that it is one of great complexity. In experiments involving the transplantation of portions of an embryo to other sites, it has been shown that some groups of cells are capable of affecting and 'organizing' the surrounding cells. Although numerous factors are probably involved, there is some evidence to suggest that this effect is due to the diffusion of specific chemical substances from the transplanted cells. The nature and pattern of these organizers varies throughout development, and they are probably responsible in part for the control of embryonal growth. An analogous situation exists in plants, where it has been known for some time that control of growth is partially mediated by growth hormones. These are specific and defined chemical substances, and have been named auxins. Some are relatively simple organic compounds such as indoleacetic acid, and they have been synthesized.

Body Size and Organ Weight

When animals of all types reach their maximum size, linear growth is then usually arrested or becomes extremely slow. How this occurs is quite unknown; it must be the result of complex interaction of physical and chemical factors, but what these may be is not clear. Even apparently rather similar animals such as the rat and mouse show a marked difference in size, while there may even be a disparity in size between strains. Related to this is the phenomenon of size and weight of organs. Most of the major organs bear a fairly close relationship in mass to each other and to the body as a whole, but as with total body size, the mechanism by which the ratios are adjusted is obscure, and will now be discussed.

Tissue Regeneration

A common, but still remarkable, example of this phenomenon is found in the regeneration of Planaria: in some of these animals small portions when excised will grow (especially if taken from the head end) into an entire animal, morphologically indistinguishable from the donor. Such complete regenerative power, akin to embryological development, is lost in the higher animals, but obviously reparative processes must occur in mammals, and indeed the life of some cells is normally very short. Some illustrative examples will be discussed.

It has been realized for some time that the number of red cells circulating rises in chronic anoxic states, and it is now definitely known that a humoral factor is released into the blood when the oxygen tension is low. This is named erythropoietin; it appears to be formed mainly in the kidney, and it stimulates erythropoiesis in the marrow, thus providing at least one mechanism for the control of the red cell population. However, the situation is probably more complex, because not only is it produced by several tissues, but other forms occur; also, we do not know its precise mode or site of action, and still less do we know what controls the actual release of newly formed red cells into the blood.

The mammalian liver and kidney also show the phenomenon of regeneration (or, more accurately, of restoration) after loss of part of the functioning cell mass. Thus, if one kidney is removed, there is hypertrophy of the remaining kidney until its mass is equal to that of the original two organs. No formation of new nephrons occurs, but the functional capacity of those remaining increases, and mitotic divisions can be seen in the tubule cells of the remaining kidney (Rollason, 1949). Similarly, when two-thirds of the liver is removed surgically in the rat, the remaining one-third will grow back to its full weight, and is histologically and functionally normal, within ten days. This restoration can also follow acute loss of tissue due to the action of toxins such as carbon tetrachloride.

The main changes in the liver of the rat are shown in *Figure 8.7*, drawn from the data of Grisham and other authors (Grisham, 1962). The disparity between

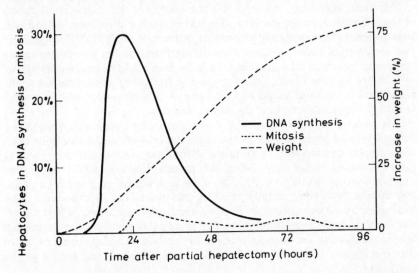

Figure 8.7. The changes in the rat liver following partial hepatectomy

the percentage of cells undergoing DNA synthesis and mitosis indicates that the hepatocytes remain longer in the synthetic, than in the mitotic, phase. The mean peak mitotic rate of almost 4 per cent is in fact very striking if it is realized that the normal mitotic rate is 0·0001 per cent (Bucher, 1967).

The regenerative power of the liver and kidney has attracted much attention, and several theories have been advanced to explain this type of tissue homoeostasis. It is evident that certain general physiological factors influence tissue hypertrophy and hyperplasia, such as nutrition, temperature, blood supply, the main endocrine hormones and general metabolic requirements, to name only a few (Paschkis, 1958). For example, feeding protein will produce a mitotic wave in the hepatocytes, and over a period of time can induce an increase in cell mass.

317

In the instances quoted above, it has been suggested that 'work hypertrophy' is responsible for the changes. It is known, for example, that after repeated exercise there is an increase in the amount of contractile protein in a muscle (but not in the number of muscle fibres). Similarly it is argued, the hyperplasia of kidney and liver is triggered by the extra work-load imposed. Proponents of the theory offer little evidence for this view; in the case of the kidney it is unlikely to be urea, for instance, as the blood levels rise only briefly and moderately in the first few hours after unilateral nephrectomy, whereas the mitotic response continues for some 10–14 days. Furthermore, if the work-load is increased by allowing one ureter to drain into the peritoneal cavity, there is no stimulation of growth in the contralateral kidney (Royce, 1963).

For these, and other, reasons it has been suggested that humoral substances of some specific type are involved. One widely quoted hypothesis, which has at least the merit of simplicity and plausability, is due to Weiss (1955). He considered that each type of tissue produced a specific 'anti-template' to growth-stimulating 'templates' present in the same tissue. Thus the liver would secrete 'anti-template' or inhibitor into the blood, and in re-entering the cell by some other route, the blood level of inhibitor would be monitored. A loss of tissue would cause a fall in blood levels, and thus trigger the mitotic response.

The mitotic curve in *Figure 8.7* shows a secondary peak which suggests a damped type of response such as one might expect with a system of this kind. Also, despite considerable experimental difficulties, some evidence is emerging that a humoral factor may indeed bé present (Bucher, 1967). Again, the situation must be multifactorial, as the increase in weight, nuclear DNA content and mitotic rate, while normally moving *pari passu*, can be dissociated from each other by various chemicals. Further, localized regeneration of hepatocytes can occur in small traumatized areas, and presumably some other mechanism is implicated. Similar considerations apply to the kidney; and it should be noted that specific stimulatory factors have also been suggested, and cannot yet be ruled out.

Wound Healing

This is obviously one aspect of the problems presented above, and it has been studied especially in the skin (Montagna and Billingham, 1964). In the normal skin there is always some mitotic activity in the basal layer of the epidermis, and the dividing cells tend to be present in small clusters as though influenced by some local factors. When an incision is made, the cells at the wound margin show marked mitotic activity, and tend to spread across the gap to form a complete layer. Cells in tube culture spread in this way and cease to do so, or to divide, when a single continuous sheet (a monolayer) has been formed. This is called 'contact inhibition' and it has been suggested that in some way the competitive adhesion of cell membranes is a factor in mitotic homoeostasis

(Carter, 1968), but there is no evidence as to how this might operate. Such a mechanism might be true for epithelial tissues, for example, but it is unlikely to be the only factor. The fact that cortisone will inhibit mitosis in normal, but not in wounded, skin (Davis, 1964), suggests that control mechanisms can vary with alterations in the state of the tissue.

Suggestions have been made that specific wound hormones are released, but there is still little evidence for these, although many growth-promoting factors and fractions have been described, mostly acting in the rather unreal conditions of tissue culture. The theory put forward by Weiss has been extended to cover mitotic homoeostasis in skin, and the term chalone has been revived (as an antonym to hormone) implying that an inhibitory substance is normally present and is removed on wounding (Bullough and Rytŏmaa, 1965).

It will be evident that the control of growth in a tissue in general, and of mitosis in particular, is of great complexity, and the factors and mechanisms concerned probably vary enormously from one tissue to another. The whole subject is still largely unexplored; for a general discussion see Swann (1957). Some interesting aspects and theories are discussed in a recent symposium (Wolstenholme and Knight, 1969).

WATER, ELECTROLYTE AND ACID-BASE BALANCE

Water

Some 60–70 per cent of the body weight is due to water, and it is clear that the control of total content of body water must be of importance in general homoeostasis. Its regulation is still not understood in detail; in mammals, at least, it is largely related to the sodium concentration of the extracellular fluid (ECF), and in fact osmolality is controlled more closely than total fluid volume. The water content of an individual cell depends largely on the difference between the internal and external osmotic pressures, with the ECF osmolality being an important determining factor. The role of sodium is considered in the next section, but two major mechanisms exist for the control of water intake and output.

Thirst. – This appears to be related to an increase in the osmolality of body fluid, rather than to a decrease in its volume. It is probably mediated by cellular dehydration, possibly affecting specialized groups of cells, and these may act through a hypothalamic 'thirst' centre.

Anti-diuretic hormone (ADH). – The kidney is the main agent in the control of water balance. Water may be lost *via* the kidney in an *osmotic* diuresis, when a large amount of solute (such as urea or glucose) has to be excreted, and the kidney can exert very little regulatory action on this loss. On the other hand, water may be lost as a *water* diuresis, (that is, as free water), and the urine is of low specific gravity, often close to that of pure water.

This water loss is controlled by the secretion of ADH, which is stored as granules in the posterior lobe of the pituitary gland, and which is released under the direct influence of neural impulses from the hypothalamus. It was shown long ago by Verney that a rise of 2 per cent or more in the osmolality of the blood perfusing the internal carotid system, would cause a release of ADH. This is a sensitive mechanism responding rapidly to acute changes and is probably mediated by specific osmoreceptors in the walls of the supra-optic nucleus of the hypothalamus. A more sluggish response is due to baroceptors in the right ventricle and the carotid sinus, which respond to changes in mean pressure and thus indirectly to blood volume.

It is also known that stress, and drugs such as morphine, will stimulate ADH secretion; a syndrome of 'inappropriate secretion of ADH' following operation, injury and other acute illnesses, is well recognized (Bartter and Schwartz, 1967). Water is retained, and a fall in the concentration of all the plasma components occurs.

The major action of ADH appears to be on the cells lining the collecting tubules. These tubules pass through a zone of high osmolality in the renal papillae and, if the cells are permeable to water, concentration of the urine occurs due to osmotic equilibration with the interstitial fluids (that is, passive transport). Obviously, concentration beyond the osmolality of the interstitial fluid is not possible. The permeability of the collecting duct cells is regulated by ADH; if the hormone is absent, then no water can be transferred by osmosis, and a dilute urine, equal in specific gravity to that leaving the distal convoluted tubule, is excreted.

In the disease of diabetes insipidus, which follows complete loss of ADH-producing tissue, the patient may pass up to 30 litres of urine a day, the body cells being permanently in danger of dehydration. A familial nephrogenic form also occurs in which the pituitary secretes ADH, but the cells of the collecting tubules fail to respond, again resulting in a massive diuresis.

The action of ADH is purely on the passive reabsorption of water; however, the suggestion has been made that the kidney is able to handle a water load not only by inhibition of ADH secretion, but by the active tubular secretion of free water (Dormandy, Robinson and Wright, 1967).

Sodium

The homoeostatic regulation of sodium is far from being thoroughly understood, as many physiological factors are involved. The individual cell contains relatively little sodium (10 mEq/l as compared to 140 mEq/l for the ECF) and maintains this differential by the active energy-consuming mechanism of the sodium pump. There is evidence to suggest that this pump may be partly controlled by adrenocortical steroids. The ECF concentration of sodium is important for the integrity of individual cells, and therefore in higher animals it is carefully regulated, solely by the action of the kidney.

The glomerular transudate passes down the proximal tubule and some 80—90 per cent of sodium is removed by a process of active reabsorption which is incompletely understood; however there is strong evidence that it is under the control of a humoral inhibitor, linked to the volume of the ECF (Editorial, 1967b). In the ascending limb of Henle, sodium is actively reabsorbed, and this is further achieved in the distal tubule where it is transported by a specific carrier which appears to be mainly under the control of aldosterone.

Aldosterone is a steroid formed in the adrenal cortex, probably entirely in the zona glomerulosa; when secreted it induces the formation of the specific transport protein by an RNA-mediated mechanism. It thus takes several hours to produce its effect. ACTH has only a minor effect on the release of aldosterone, and the main agent appears to be angiotensin II. This is an octapeptide derived from a decapeptide angiotensin I which in turn is derived from the action of the enzyme renin on an α_2-globulin substrate in the plasma. Renin is present in granule form in the juxtaglomerular apparatus of the kidney, and its release appears to be stimulated by a fall in the pressure in the afferent glomerular arteriole. By the above mechanism this produces sodium retention with expansion of the ECF volume and a rise in blood pressure. Renin release may perhaps also follow a fall in the concentration of sodium in the distal tubule. The kidney thus reabsorbs 99 per cent of filtered sodium, and exerts fine control in the distal tubule.

Some doubt has been cast on the importance of aldosterone, especially as the tubule cells appear to be able to 'escape' from its sustained action after several weeks of experimental administration; it is thought that cortisol may be at least as important. With complete loss of adrenal tissue, as in Addison's disease, sodium loss is extreme unless steroids are given, and control can in fact be achieved with a glucocorticoid such as cortisol. However, in pituitary necrosis and after hypophysectomy, there is much evidence to suggest that aldosterone levels remain normal (Cope, 1964), despite a marked fall in cortisol output: although such patients may go without treatment for many years they rarely have severe electrolyte problems. Thus both cortisol and aldosterone must have a role in sodium regulation.

Other factors such as some form of hypothalamic control, and possibly a salt-retaining steroid, may be involved, but there is still no unequivocal evidence available; these aspects are discussed by Cope (1964).

Acid—base Balance

The hydrogen ion (hydrion) content of normal plasma is maintained within the range pH 7·44—7·36 (36—44 nanoequivalents H^+/l), and intracellular pH is also carefully guarded. Some 60 mEq H^+/day are formed in normal metabolism, and have to be excreted; this amount corresponds to one-tenth of the maximum buffering capacity of the body.

The hydrogen ion in the ECF is buffered chiefly by the bicarbonate system, and by protein (mainly haemoglobin), the actual pH depending, as in all buffer systems, on the ratio of the components rather than their concentration. Thus the following generalization of the Henderson—Hasselbalch equation is useful:

$$[H^+] \; \alpha \; [\text{metabolic component}] / [\text{respiratory component}]$$

Respiratory Component

This component is H_2CO_3 in the Henderson equation, and is directly proportional to P_{CO_2} (in fact, because of the reaction equilibrium, very little carbonic acid is actually present). The carbon dioxide—carbonic acid concentration is effectively adjusted by alterations in the ventilation rate, which is under the control of the respiratory centre. The fluctuations in P_{CO_2} and pH are monitored by a complex system of chemoreceptors, and by the respiratory centre itself. Rapid adjustments can be made in the ventilation rate (and thus in the blood P_{CO_2}) to compensate for changes in the metabolic component.

Metabolic Component

This is represented by total buffer base: chiefly HCO_3' and haemoglobin, with a minor contribution from other proteins and phosphates. Where the change in respiratory component is due to a primary disturbance of P_{CO_2} (for example, in hyperventilation), secondary adjustments are made to the metabolic component by the kidney until the primary condition is corrected. If there is a primary change in the metabolic balance of H^+ (due, for example, to formation of ketone bodies in diabetes mellitus) then, although the change in the metabolic component is rapidly compensated for by alteration in ventilation rate, the imbalance has ultimately to be corrected by the kidney.

The kidney is, in fact, the sole organ concerned with the regulation of hydrogen ion excretion, and this regulation occurs in two main ways:

(1) H^+ and HCO_3' are generated in the epithelial cells of the proximal segment of the renal tubule from carbonic acid by the action of carbonic anhydrase (*Figure 8.8*). Due to differential membrane permeability the H^+ diffuses into the lumen of the tubule, while HCO_3' diffuses into the blood, thus restoring buffer base. Once the hydrogen ion is in the tubule it may be disposed of in four ways:

(a)   it will combine with any bicarbonate in the glomerular transudate, forming water and carbon dioxide, which will be reabsorbed. Thus base is conserved.

322

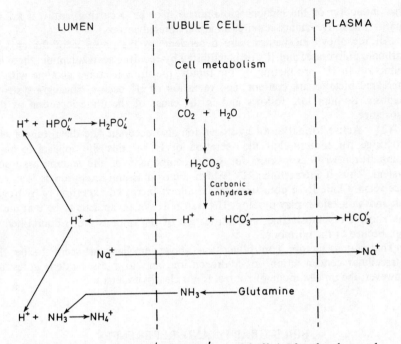

Figure 8.8. Generation of H^+ and HCO'_3, and buffering by phosphate and ammonia, in the renal tubule

(b)   it can be excreted as free H^+; the amount handled in this way is extremely small (approximately 0·5 per cent of the average daily load, at pH 4).

(c)   the major mechanism for the disposal of normal acid loads is buffering with phosphate:

$$H^+ + Na_2HPO_4 \longrightarrow NaH_2PO_4 + Na^+$$

Thus the hydrogen ion generated is exchanged for sodium, and the latter is reabsorbed.

(d)   the hydrogen ion may combine with ammonia, which is actively formed in the cells of the distal segment, chiefly from glutamine by the action of glutaminase. The ammonia diffuses readily through the wall into the lumen and forms NH_4^+ with the hydrogen ion. The cell wall is relatively impermeable to NH_4^+ and therefore back diffusion does not occur.

This is quantitatively far the most important mechanism when a large load of hydrogen ion has to be excreted. The formation of ammonia shows a lag response to a sustained acid load, and rises slowly over several days; this suggests that induction of the enzyme glutaminase may be occurring, and it is known that this adaptive change requires an intact adrenal cortex.

All the above mechanisms are dependent on the presence of the enzyme carbonic anhydrase, and if this is inhibited by the drug acetazolamide, there is a reduction in H^+ production by the tubule, the urine becomes alkaline with an increased bicarbonate content, and retention of H^+ occurs, causing a systemic acidosis. Sodium loss follows and is the cause of the diuretic action of this substance.

(2) Active formation of hydrogen ion also occurs in the distal tubule with exchange for sodium, but the secretion of H^+ at this site appears to be in competition with potassium for some component of the sodium transport system. Thus if intracellular H^+ is high then potassium excretion is low, and vice versa. Clinically a potassium deficit allows increased excretion of hydrogen ion and an alkalosis may develop, The transport system appears to be that under the control of aldosterone, as acidification of the urine is much diminished in the absence of the hormone.

The actual factors controlling the above mechanisms seem to be the intracellular concentration of hydrogen ion, and to a lesser extent potassium. However, the precise mode of control is not clearly understood.

BIBLIOGRAPHY AND REFERENCES

Bartter, F. C. and Schwartz, W. B. (1967). 'Syndrome of Inappropriate Secretion of ADH.' *Am. J. Med.* **42**, 790.

Black, D. A. K. (1964). *Essentials of Fluid Balance*, 3rd ed. Oxford; Blackwell.

Bucher, N. L. R. (1967). 'Experimental Aspects of Hepatic Regeneration.' *New Engl. J. Med.*, **277**, 686, 738.

Bullough, W. S. and Rytömaa, T. (1965). 'Mitotic Homeostasis.' *Nature, Lond.*, **205**, 573.

Burns, T. W., Hales, C. N. and Hartree, A. (1966). 'Lipolytic Activity of Human Pituitary Fractions on Isolated Adipose Tissue Cells.' *Lancet*, **2**, 1111.

Carter, S. B. (1968). 'Tissue Homeostasis and the Biological Basis of Cancer.' *Nature, Lond.*, **220**, 970.

Clegg, P. C. and Clegg, A. G. (1969). *Hormones, Cells and Organisms*. London; Heinemann.

Cope, C. L. (1964). *Adrenal Steroids and Disease*. London; Pitman.

Davenport, H. W. (1958). *The ABC of Acid—Base Chemistry*, 4th ed. London; University of Chicago Press.

Davidson, J. N. (1965). *The Biochemistry of the Nucleic Acids*, 5th ed. London; Methuen.

Davis, J. C. (1964). 'The Effect of Cortisol on Mitosis in the Skin of the Mouse.' *J. Path. Bact.*, **88**, 247.

Dormandy, T. L., Robinson, A. and Wright, E. D. (1967). 'Active Renal Water Excretion.' *Lancet*, **1**, 246.

Editorial (1967a). 'Human Growth Hormone and Insulin.' *Lancet*, **1**, 257.

— (1967b). 'The Fate of Sodium in the Renal Tubules.' *New Engl. J. Med.* **279**, 381.

— (1968). 'Prostaglandins.' *Lancet*, **1**, 30.

Granick, S. (1966). 'The Induction *in vitro* of the Synthesis of Delta-aminolevulinic Acid in Chemical Porphyria.' *J. biol. Chem.*, **241**, 1359.

Greenwood, F. C. and Landon, J. (1966). 'Assessment of Hypothalamic-pituitary Function in Endocrine Disease.' *J. clin Path.*, **19**, 284.

Grisham, J. W. (1962). 'A Morphologic Study of DNA Synthesis and Cell Proliferation in Regenerating Rat Liver.' *Cancer Res.*, **22**, 842.

Harper, H. A. (1967). *Review of Physiological Chemistry*, 11th ed. Altos; Lange Medical Publ.

Hirsh, J. and Gallian, E. (1968). 'Methods for the Determination of Adipose Tissue Cell Size in Man and Animals.' *J. Lipid Res.*, **9**, 110.

Kaldor, G. (1969). *Physiological Chemistry of Proteins and Nucleic Acids in Mammals.* Philadelphia; Saunders.

Masoro, E. J. (1968). *Physiological Chemistry of Lipids in Mammals.* Philadelphia; Saunders.

Montagna, W. and Billingham, R. E. (Eds.) (1964). *Wound Healing. Advances in Biology of Skin*, vol V. Oxford; Pergamon Press.

McGowan, G. K. and Walters, G. (Eds.) (1969). Disorders of Carbohydrate Metabolism. Supplement to *J. clin. Path.*, **22** (Suppl.).

Paschkis, K. E. (1958). 'Growth Promoting Factors in Tissues: A Review.' *Cancer Res.*, **18**, 981.

Rollason, H. D. (1949). 'Compensatory Hypertrophy of the Kidney of the Young Rat.' *Anat. Rec.*, **104**, 263.

Royce, P. C. (1963). 'Inhibition of Renal Growth Following Unilateral Nephrectomy in the Rat.' *Proc. Soc. exp. Biol. Med.*, **113**, 1046.

Samols, E., Marri, G. and Marks, V. (1965). 'Promotion of Insulin Secretion by Glucagon.' *Lancet*, **2**, 415.

Swann, M. M. (1957). 'The Control of Cell Division: A Review.' *Cancer Res.* **17**, 727.

— (1958). *Cancer Res.*, **18**, 1118.

Symposium (1968). 'Symposium on Glycogen and Glycogen Deposition Diseases.' *Am. J. clin. Path.*, **50**, 1–51.

Tata, J. R. (1969). 'Hormonal Control of Protein Synthesis.' In *Scientific Basis of Medicine, Annual Reviews*, chapter 7. British Postgraduate Medical Federation. London; Athlone Press.

Vance, J. E., Buchanan, K. D. and Williams, R. H. (1968). 'Effect of Starvation and Refeeding on Serum Immunoreactive Glucagon and Insulin Levels.' *J. Lab. clin. Med.*, 72, 290.

Weiss, P. (1955). 'Specificity in Growth Control.' In *Biological Specificity and Growth*. Ed by E. G. Butler. Princeton; Princeton University Press.

White, A., Handler, P. and Smith, E. L. (1968). *Principles of Biochemistry*, 4th ed. New York; McGraw-Hill.

Wolstenholme, G. E. W. and Knight, J. (Eds.) (1969). 'Homeostatic Regulators.' Ciba Foundation Symposium. London; Churchill.

Zahnd, G. R., Nadeau, A. and von Muhlendahl, K. E. (1969). 'Effect of Corticotrophin on Plasma Levels of Human Growth Hormone.' *Lancet*, 2, 1278.

Index

Guanosine triphosphate (GTP) in
protein synthesis, 208

Haem, 68, 91
Haemochromes, 91
Haemoglobin
    amino-acid substitution in, 222,
        223
    categories, 70
    chains, 68, 69, 223
    haem portion, 68
    oxygenation of, 70
    sickle-cell disease, in, 222
    structure, 67, 68
Haemophilia, gene responsible for, 186
Hand, prehensile, 16
Haploidy, 187
Haptens, 230
Hardy—Weinberg equilibrium, 185
Harvey, William, 28
Heart, evolution of, 13
Heat, conversion of, 121
Heparin, 82
Herbivores, evolution, 12
Hereditary angioneurotic oedema, 264
Hereditary thymic aplasia, 255
Hexose, degradation, 138
Hexose shunt, 303
Histamine, 266, 267
Histidine, detection, 57
Histones, 60
Holocene period, 14
Hominids, culture of, 17
    definition of, 15
    evolution of, 18, 19—24
Hominoids, definition of, 15
Homoeostasis, 296, 305—306
    carbohydrate, 303
    glucose, 305
    sodium, of, 320
Homo erectus, 18, 20
Homogentisic acid, 219
Homografts, survival, 251
Homo sapiens neanderthalensis, 18, 20
Homo sapiens sapiens, 18, 21
Hooke, Robert, 28

Hormones, 295—297
    carbohydrate metabolism, and, 303
    production of, 295
    protein synthesis, in, 312
Host—parasite relationship, 280
Host versus graft response, 252
Humanists, 27
Hunting, 22
Hyaluronic acid, 82
Hydrogen, reaction with oxygen, 124
Hydrogen ion concentration, 45, 162,
    321
Hypergammaglobulinaemias, 256
Hypertrophy, work, 318
Hypophosphatasia, 218
Hypothalamus
    hormones and, 285
    sodium regulation by, 321

Ice ages, 14
Immune response, suppression of, 253
Immunity, 227
Immunoelectrophoresis, 233
Immunoglobulins
    antibodies associated with, 241
    antigenic determinant sites, 235
    classification and properties, 240
    disordered synthesis, 254
    IgA,
        antibodies associated with, 241
        formation of, 250
        normal, 241
        properties, 240
    IgD, 234
        normal, 242
    IgE, 234
        normal, 242
        properties, 240
    IgG, 234
        antibodies associated with, 241
        antibody reaction sites, 235
        chromatography, 234
        formation, 250
        molecular model, 235—238
        normal, 239
        properties, 240

Light, 147, 150, 151
Limbs in primates, 16
Linkage and crossing-over, 180
  map, 182
Lipids, 83—86
  cell membrane, in, 39, 100
  cerebral, 86
  classification, 84
  complex, 85
  metabolism,
    diabetes mellitus and, 310
    regulation of, 306
  simple, 84
Lipoproteins, 61
Liver
  cells, 104
  regeneration, 316
Lorisoidea, 14, 15
Luciferin, 155
Lutein, 91
Luteinizing hormone (LH), 296
Lymph nodes
  defence system, as, 282
  immunogens in, 244
  plasma cells in, 245
Lymphocytes
  antibody formation and, 244, 247
  metamorphosis, 246
  migratory routes, 249
  types of, 245
Lysosomes, 111, 112

Macrophages, 112, 282
  antibody formation, in, 247
Magdalenians, 22
Magnesium, complex with adenosine
    triphosphate, 128, 129
Maleylacetoacetic acid, 219
Malic acid, 141
Malphigi, Marcello, 28
Maltose, 81
Mammals
  age of, 12
  evolution of, 12
  reproduction in, 13
Mammary glands, 13

Man
  evolution of, 17, 18, 19—24
  future of, 24
  mechanistic view of, 24
Marsupials, evolution of, 12
Mass action, law of, 125, 290
Meiosis, 176
  anaphase I, 178
  anaphase II, 179
  crossing-over in, 180
  division I, 177
  division II, 178
  metaphase I, 177
  metaphase II, 179
  prophase I, 177
  prophase II, 178
Mendel's first law of genetics, 179
Mesozoic era, 9, 11
Metabolism, 5
  control of, 289
  generation of adenosine phosphates
    in, 132
  inborn errors of, 217
Metatheria, 12, 13
Methaemoglobinaemia, 218
Michaelis Menten constant, 46
Microbes, discovery of, 29
Microphagocytosis, 113
Microscope, 28
  achromatic lenses, 29
  dark-ground illumination, 33
  definition of, 31
  electron, 33
  evolution of, 28
  magnification of, 31
  phase-contrast, 32
  theory of, 30
Miocene era, 14, 17, 18
Mitochondria, 38, 105
  aerobic respiration and, 143
  membrane, 99, 107
  morphology, 106
Mitosis, 174
  anaphase, 175
  discovery of, 30
  metaphase, 175
  prophase, 175

Proteins, *cont*
  antigens, as, 229
  cell membranes, in, 39, 100, 103
  characteristics and classification, 57
  chromatography, 59
  conjugated, 61
  crystallization of, 60
  disulphide bonds, 65
  electrophoresis of, 60
  isoelectric precipitation, 59
  molecular weight, 58, 61
  osmotic pressure, 58
  peptide bonds, 62
  polypeptide chains, 64
  purification of, 59
  relation to nucleic acids, 202
  salt precipitation of, 59
  simple, 60
  solubility of, 58
  specificity, 58
  stability, 58
  structural organization of, 61
    primary, 63
    secondary, 65
    tertiary and quaternary, 67
  synthesis,
    control of, 70, 208–210
    endoplasmic reticulum and, 104
    hormonal role in, 312
    Jacob and Monod theory, 209
    process of, 205
    regulation of, 311
    ribonucleic acid in, 75
    ribosomes, role of, 38
Proterozoic era, 9
Protheria mammals, 13
Protista, 40
Protoculture, 18
Protophyta, 40
Protoplasts, 95
Protozoa, 40
Prozone, 261
Purines, 70
Pyrimidines, 70
Pyrimidine nucleosides, synthesis, 313
Pyruvate, 298, 302
Pyruvate phosphokinase, 138

Pyruvic acid
  conversion to acetyl coenzyme A, 140
  de-carboxylation of, 135
  fate of, 139
  formation of, 133
Pyruvic dehydrogenase complex, 38

Quantum theory, 149
Quaternary period, 14

Radiation, mutation induced by, 189
Ramapithecus, 17
Rathke's pouch, 296
Re-combination in genetics, 181
Red cells (*See* Erythrocytes)
Reduction, 130
Renaissance, scientific development in, 27
Reproduction in mammals, 13
Reptiles, 11
Respiration
  acid–base balance and, 322
  aerobic, 140
  anaerobic, 132–140
  bacterial, 132
Respiratory chain, 144, 146
  citric acid cycle and, 145, 146
  coupling reaction, 145
Reticulo-endothelial system, 231
  body defence by, 282
Rh blood groups, 279
Rhodopsin, 30
Riboflavine, 92
Ribonuclease, 64, 66
Ribonucleic acid (RNA), 70
  composition, variation of, 74
  forms of, 74
  nucleus, in, 37
  structure of, 73, 74
  synthesis,
    antibody formation and, 247
    hormonal role, 312
mRibonucleic acid (mRNA), 74, 311
  formation of, 75
  function in protein synthesis, 205, 209